SELLING ETHNICITY

Heritage, Culture and Identity

Series Editor: Brian Graham,
School of Environmental Sciences, University of Ulster, UK

Selling EthniCity
Urban Cultural Politics in the Americas

Edited by
OLAF KALTMEIER
Bielefeld University, Germany

ASHGATE

Published by
Ashgate Publishing Limited
Wey Court East
Union Road
Farnham
Surrey, GU9 7PT
England

Ashgate Publishing Company
Suite 420
101 Cherry Street
Burlington
VT 05401-4405
USA

www.ashgate.com

British Library Cataloguing in Publication Data
Selling ethnicity:urban cultural politics in the Americas. – (Heritage, culture and identity)
 1. Ethnicity–Political aspects–America. 2. Ethnicity–Economic aspects–America.
 3. Politics and culture–America. 4. Indigenous peoples in popular culture– America.
 5. Multiculturalism in mass media. 6. America– Ethnic relations.
 I. Series II. Kaltmeier, Olaf, 1970–
 305.8'0097-dc22

Library of Congress Cataloging-in-Publication Data
Kaltmeier, Olaf, 1970–
 Selling ethnicity : urban cultural politics in the Americas / by Olaf Kaltmeier.
  p. cm. — (Heritage, culture, and indentity)
 Includes bibliographical references and index.
 ISBN 978-1-4094-1037-9 (hardback) — ISBN 978-1-4094-1038-6
 (ebook) 1. Ethnicity—Political aspects—America. 2. Ethnicity—Economic aspects—America. 3. Politics and culture—America. 4. Indigenous peoples in popular culture—America. 5. Multiculturalism in mass media—America. 6. America—Ethnic relations.
 I. Title.
 GN550.K35 2011
 305.800970—dc22

2010045835

ISBN 9781409410379 (hbk)
ISBN 9781409410386 (ebk)

MIX
Paper from
responsible sources
FSC® C018575
www.fsc.org

Printed and bound in Great Britain by the
MPG Books Group, UK

Contents

List of Figures and Tables

Figures

Tables

Notes on Contributors

Selma Siew Li Bidlingmaier holds a position as Lecturer in American Studies at the Ruhr-University Bochum, Germany. She is also concurrently completing her PhD dissertation on the writing of Chinatowns as third spaces.

Martin Butler is Junior Professor of American Literature and Culture at the University of Oldenburg, Germany.

Fernando Carrión Mena is coordinator of the *Programa Estudios de la Ciudad* of the FLACSO Ecuador. He is president of the *Organización Latinoamericana y Caribeña de Centros Históricos* and author and editor of numerous articles and books on urban studies in the Americas.

Manuel Dammert Guardia holds a professorship at the Pontificia Universidad Católica del Perú, Lima. He also serves as the editor of the Urban Studies journal *Centro–h*.

Eric C. Erbacher is a Ph.D. candidate in American Studies at the Technical University Dresden, Germany. He has held positions as research assistant and lecturer in American Studies at the universities of Dresden, Bonn, and Paderborn, Germany.

John R. Gold is Professor of Urban Historical Geography in the School of Social Sciences and Law at Oxford Brookes University. A frequent radio and television broadcaster, he has authored or edited numerous books on architectural and cultural subjects.

Jens Martin Gurr holds the Chair in British and Anglophone Literature and Culture at the University of Duisburg–Essen, Germany. He currently also serves as Speaker of the university's main research area "Urban Systems."

Olaf Kaltmeier is Junior Professor of Transnational History of the Americas at the Department of History at Bielefeld University, Germany. He is co-organizer of the Research Group "E Pluribus Unum? Ethnic Identities in Transnational Integration Processes in the Americas" at the Center for Interdisciplinary Research and Executive Director of the Center for InterAmerican Studies at Bielefeld University.

Jens Kastner, Dr. phil., a sociologist and art historian, works as Research Associate at the Institute for Art Theory and Cultural Studies at the Academy of Fine Arts, Vienna.

Ricardo López Santillán holds a Ph.D. in Sociology from the Université de Paris III–Sorbonne Nouvelle. He works as Researcher at the Peninsular Center for Humanities and Social Sciences of the Universidad Nacional Autónoma de México in Mérida de Yucatán.

Alicia Menéndez Tarrazo is a Ph.D. candidate in English Language and Literature at the University of Oviedo, Spain, where she also teaches Postcolonial Literatures in English.

Nina Möllers holds a Ph.D. in History and works as researcher in the project "Objects of Energy Consumption" at the Research Institute of the Deutsches Museum, Munich.

Ruxandra Rădulescu holds a position as Junior Lecturer in the Department of English at the University of Bucharest.

Wilfried Raussert is Professor of North American Literary and Cultural Studies at Bielefeld University, Germany. He is general editor of the journal *Fiar: Forum for Inter-American Research* as well as Executive Director of the International Association of Inter-American Studies.

Christina Seeliger is a Ph.D. candidate and Lecturer in English and American Studies at Bielefeld University, Germany.

Juliana Ströbele-Gregor, Dr. phil., is a cultural and social anthropologist and holds a position as Associated Researcher at the Latin American Institute of the Free University Berlin.

Julie TelRav is a Ph.D. candidate in Sociology at the New School for Social Research in New York City. She is currently teaching at Johnson and Wales University in Denver, Colorado.

Ulises Zarazúa Villaseñor is Professor at the Socio-Urban Studies Department at the University of Guadalajara, Mexico. He is also known as a writer, especially on urban imageries.

Acknowledgments

The volume *Selling EthniCity* has emerged out of the Research Group "E Pluribus Unum? Ethnic Identities in Transnational Integration Processes in the Americas" (2008–2010), which was organized by Sebastian Thies, Josef Raab, and Olaf Kaltmeier at the Center for Interdisciplinary Research (ZiF) at Bielefeld University, Germany. My first thanks must go to all of the authors whose work is included in this volume, with whom we had fruitful discussions at the sessions and conferences of the Research Group, and who have been vividly engaged in shaping this volume.

I am especially grateful to Astrid Haas who did a great job in the editing process from the formatting of the texts via the corrections to conceptual ideas to rethink the structure of the book. Alethea Wait, Luisa Ellermeier, and Daniela Opitz were also very much involved in the production of the book. The Center for Interdisciplinary Research offered us all the infrastructural and moral support to realize this project, and we would like to thank especially Britta Padberg, the Executive Secretary of the ZiF. I also would like to express my gratitude to Valerie Rose, Bethan Dixon, Sarah Horsley, Carolyn Court and Jude Chillman from Ashgate Publishing as well as to the editors of the "Heritage, Culture and Identity" series for their confidence in this publication project.

Introduction
Selling EthniCity: Urban Cultural Politics in the Americas from the Conquest to Contemporary Consumer Societies

Olaf Kaltmeier

> Vancouver 2010's graphic identity began with the selection of the Olympic Games emblem, Ilanaaq ["friend" in the Inuit Inuktitut language] and the development of the Paralympic Games emblem. Vancouver 2010 graphics use colors and shapes that highlight the breathtaking coast, forests, and mountain peaks in the host region. Abstract urban graphics and digitally-inspired elements represent Canada's modern cities and leading-edge technology and innovation. (http://www.vancouver2010. com/more-2010-information/about-vanoc/the-vancouver-2010-brand/).

In thus describing its corporate design, the Vancouver Organizing Committee (VanOC) reveals the marketing strategy of the city of Vancouver for the 2010 Olympic Winter and Paralympic Games. The combination of place promotion and strategic use of ethnicity in city-marketing are outstanding, as is the fact that these were the first Olympics in which the International Olympic Committee recognized indigenous nations—the four Canadian First Nations of the Vancouver region—as official hosts and partners. Gary Youngman, consulting director for aboriginal participation, pointed out to what extent indigeneity became a unique symbol for the Vancouver Olympics:

> One of our greatest challenges is that Indigenous participation is relatively new to the Olympic Movement—there is no template we can follow—no clear indicators for how we measure our success. Indigenous participation in past Games, such as Calgary and Salt Lake City, has focused primarily on ceremonies and cultural programs. We plan to go beyond that, to set the bar higher, with the hope that future Organizing Committees can be inspired and learn from our experience. (http://www.vancouver2010.com/aboriginal-participation/)

Nevertheless, the use of ethnicity did not remain uncontested by the First Nations. One point of contention was the mascot for the Games: much to the annoyance of indigenous groups from the Vancouver region, the design of the mascot, made by an Argentinean graphic designer, did not invoke their own native culture, the Salish, but that of the Inuit, the indigenous population of the Arctic region far North from Vancouver.

However, the Vancouver example points out a general tendency in urban cultural politics. Local governments regard the utilization of culture for purposes of city marketing and image creation as a significant redevelopment and marketing strategy (Yúdice 2003, Short and Kim 1999, Zukin 1995, Gotham 2007, Kearns and Philo 1993, Gold and Ward 1994). Obviously these tendencies are related to questions of urban governmentality, urban imaginaries, and socio-spatial segregation. The questions of what and who is to be included, represented, seen, and listened to remain at stake in these negotiations. In light of these questions, the main starting point for the present volume is the idea that ethnicity can be used strategically in urban cultural politics.[1] These recent processes are related to diverse urban political cultures which have been shaped by certain patterns of settlement, ethno-racial diversity and conflicts, the materialization of former programs in architecture and city planning, the path of economic development, and a city's position in the capitalist world system, to mention just a few elements.

An approach which tackles this topic from an Urban Cultural Politics perspective offers several advantages. First, this strategy allows for the application of actor-centered concepts which conceive of the city as a space of action and performativity in which individuals and groups communicate and interact with one another as well as position themselves socially and ethnically. Alternatively, the city can be understood as a system of symbols, whose integral parts are used by actors when competing for sovereignty of interpretation. The city can be thus understood as a dense space of signs and symbols in and through which individual and collective actors articulate themselves and, by means of different cultural practices, struggle to implement their principles of vision and division of the social world.

Another related advantage of the concept of Urban Cultural Politics lays in the fact that it encompasses a vast range of distinct actors, ranging from highly institutionalized and financially strong organizations, cultural producers, and social movements, to actors of the everyday world (see also the conceptualization of the field of identity politics in Thies and Kaltmeier 2009). The 2010 Winter Olympics in Vancouver provide a case in point to exemplify this idea. As stated above, one can observe how economic and political elites—such as VanOC, the Vancouver municipal government, and (trans)national business corporations—intervene in urban political culture and cultural economy by means of ethnicized cultural productions such as the "indigenous" ethnic branding of the Olympics. The International Olympic Committee has encouraged the use of ethnicity in the marketing of the Games, as in 1999 when they adopted the *Agenda 21 for Sustainable Development in Sports*, which includes an objective to strengthen the inclusion of women, youth, and indigenous peoples in the Olympic movement.

---

1   The use of the concepts of cultural politics/ political culture is inspired by the debates on the re-conceptualization of social movements in Latin America in the late 1990s (see Alvarez et al. 1998, Kaltmeier et al. 2004).

At the same time, actors from everyday life contexts are able to use the ethnic labeling of cultural production. VanOC in particular has further supported ethnic businesses and ethno-tourism. In its online Frequently Asked Questions about indigenous peoples in the Olympics, one of the first items is: "I have an Aboriginal Business. How can I get involved?" (http://www.vancouver2010.com/more-2010-information/aboriginal-participation/aboriginal-participation-faqs/). This example expresses the ambivalent position between the commodification of ethnic identities, on the one hand, and their strategic use for changing living conditions among ethnic communities and the empowerment of their members through their integration into mainstream commercial circuits, on the other.

Cultural producers are of particular importance in this context, as they are highly likely to both attract tourists and other cultural consumers and reflect critically upon ethnic conflicts. This ambiguous position is also revealed in the Vancouver Olympics. The Cultural Olympiad, a broad cultural program targeting domestic and international visitors, was an integral part of the Olympics and addressed issues of ethnicity, among others. For example, Ken Lum's "Monument for East Vancouver" takes up the graffiti East-Van logo, a symbol of Vancouver's multiethnic urban outcasts, and VanOC promoted indigenous auto-ethnographic film productions. Nevertheless, the organizers of the Cultural Olympiad explicitly limited the critical reflection and the freedom of expression of the participating artists, as their contracts prohibited any critical or negative comments on VanOC, the Olympics, or any sponsor of the Games (Schafhausen 2010: 12).

Social movements also relied on the prominent use of ethnicity in the Olympic marketing campaigns. For instance, their slogan "No Olympics on Stolen Native Land" belied the continuing impact of colonialism in British Columbia. The instances of anti-Olympic urban riots included the smashing of windows at The Bay, a shopping center operated by the Hudson Bay Company, which had both been involved in Canadian colonialism and acted as one of the sponsors of the Vancouver Olympics.

In order to explore the multiple urban cultural politics pertaining to the recent process of "selling ethniCity," the present volume focuses on the dynamics of this phenomenon in the Western hemisphere. The Americas are some of the regions with the highest degree of urbanization in the world. According to the data of UN–Habitat, in the year 2000, 77.2 per cent of the population in South America lived in urban areas. It is not only the massive urbanization processes, but also the (post)colonial condition of the societies of the Americas and the related drawing of boundaries between the "first" and "third" world which culturally, socially, politically, and economically marks the American continent. Such interactions have turned the urban and metropolitan regions of the Americas into a laboratory for the emergence and development of pluriethnic urban cultures. Against this background, the continent provides a unique range of forms of cultural expressions that offer themselves to explorations of intercultural communication and negotiation processes, to the analysis and description of phenomena of hybridization and cultural transfer, and hence foster the examination of (trans)formations of ethnic identities. Especially with regard to politics of multiculturalism and ethnic diversity, the Americas further represent a global point of reference (Kymlicka 1995, Hollinger 1995). In this respect, the present volume contributes to the emerging field of trans-regional Inter-American Area Studies (Thies and Raab 2009, Ostendorf 2002, Sadowski-Smith and Fox 2004).

In order to explore urban cultural politics within a historical perspective, this introduction distinguishes among three layers of ethnic differentiation in the city. Under European colonial rule, the cities and towns of the Americas formed an integral part of the material and imaginary colonization process, being the most important nodes of articulation between the colonies and the colonizers' mother countries. After the colonies had gained independence, their cities and towns became constitutive elements in the formation of different nation states. In many cases the cities—especially the capitals—were seen as icons of the new nations and their postcolonial Creole elites. At the same time, the structures of socio-ethnic stratification emblematized in the cities remained intact after the end of the colonial period. However, in more recent times, new dynamics can be identified in the commodification of urban ethnic identities in globalizing consumer societies.

Colonization of Space and Population: Between the Annihilation of Ethnic Difference and Transculturation

The experiences of European conquest and colonial rule profoundly informed both urbanization processes and dynamics of ethnic classification in the Americas. The cities were the centers of colonial power and expansion, and they formed the nodes of the flows of capital, goods, and peoples circulating via the Atlantic between the Americas, Europe, and Africa (Gilroy 1993). Postcolonial studies scholars have pointed out that coloniality and modernity emerged in close connection with one another. Following the macro-sociological approach of multiple modernities, Shmuel Eisenstadt argues that the institutional patterns which developed in the context of and through colonial contacts in the Americas represent the first instances of multiple modernities. Against the argumentation of traditional—Eurocentric—theories of modernization, Eisenstadt maintains that, on the basis of European patterns of culture, new, modern dynamics and interpretations emerged in the Americas, which must be seen as autonomous. He further states that "modernity and Westernization are not identical" (2002: 8), underlining the notion that European patterns of thought and action cannot be seen as the only "authentically modern" option, although they have served as the starting point for the development of alternative modernities in the Americas. In a similar vein, Walter Mignolo (2000) emphasizes that the conquest of the Americas, through a process of "Occidentalism" (similar to Edward Said's notion of Orientalism), led to the cultural and political formation of the Western hemisphere.

Thus it stands that questions of ethnic diversity and conflict are at the heart of the constitution of the Americas as a geo-cultural formation. The Peruvian social scientist Aníbal Quijano (2000) argues that the conquest of the Americas gave birth to a "coloniality of power," which can be conceived of as a social machine that transforms ethnic and cultural differences into a binary system of differentiation and value ascription, including such pairs of socially constructed opposites as modern/traditional or white/black, and thus produces a racist system of cultural classification. Furthermore, the concept includes the creation of institutions and spaces considered appropriate for this colonial classification. Without any doubt, the city is one of these spaces in the Americas, although one can observe profound differences between the Northern and the Southern parts of the hemisphere. In order to work out the dialectical constitution of modernity and coloniality in urban spaces in the Americas, with regard to the different dynamics in the Spanish empire versus the Anglophone settler colonies,[2] I will focus in particular on the use of the grid as a paradigmatic pattern in urban planning.

In Central America and in the Andes, the Spanish conquerors needed to deal with differentiated urban structures like Tenochtitlán, the capital of the Aztec Empire, and Cusco, the capital of the Andean Empire Tahuantinsuyo. Generally, the colonizers adopted the pre-Columbian urban structure of the towns, while they replaced the existing architectural structures with Spanish-style edifices. While the grid system in Spanish America is generally referred to as a colonial import which the Spanish conquerors had taken over from ancient Rome, it is worth mentioning that Tenochtitlán featured an urban grid much broader than any Spanish city of that time.

In 1573, following the famous decree in the context of the Laws of the Indies of King Phillip II, all Spanish colonial towns were required to follow the same plan, based on the *damero* (checkered) pattern of streets. The Laws specified a square or rectangular central plaza with eight principal streets running from the plaza's corners. This form shaped the colonial towns, which nowadays are referred to as colonial city centers. The Latin American grid was a strategic tool for the colonization of space, thus converting the continent into a real laboratory for urbanity. Leonardo Benevolo sites the Americas in the midst of the modern history of urban planning:

> It would be a mistake to consider the American experiments as marginal episodes in the history of architecture; they were not only quantitatively the most remarkable schemes realized in the sixteenth century, but were also in some ways the most significant, because their characters depended more upon the cultural concepts developed at this time, and less upon the resistance put up by the environment. (1978: 430)

---

2   I am aware of the fact that we can find different regional and local dynamics that can not be taken into account within the scope of this overview; this includes the particular examples of French and Portuguese colonization in the Americas.

Hence, this urban form does not represent a colonial mimesis of European examples, but a modern, utopian invention that partakes in processes of transculturation, relying also on the pre-Columbian city structures of the Aztecs and Incas. The grid was further measured and limited to the size of the ideal city. Architectonically, the religious and political dominance of the Spaniards manifested itself in monumental colonial buildings. Nevertheless, as Walter Mignolo argues, the baroque architecture in the Americas was not quite the same as its Spanish original, as it was shaped by "colonial difference" (2005: 62). Mignolo points out: "In the colonies, the Baroque was the expression of protest, complaint, rebellion, and critical consciousness by socially and economically displaced Creoles of Spanish descent. It was indeed the cry of the White Creoles feeling the pain of the colonial wound" (2005: 62). Within the urban architecture, one can therefore identify styles such as the *Barroco de Indias* or the Baroque School of Quito, an ambiguous amalgam of white supremacy and indigenous cosmovision. However, these carefully planned cities formed the centers of the colonial power in the New World, where institutional and legal power was administered through a specialized cadre of elite men called *letrados* (lettered men). Ángel Rama has called this urban nexus of intellectual culture and state power "the lettered city" (1984), which imposed a new geopolitics of knowledge.

Although the colonial city was never characterized by full ethnic segregation—indigenous markets and employees also shaped its image—a symbolic rural/urban divide (Mallon 1992) was installed. To this day, towns and cities stand for civilization and a white population, while the rural hinterland recalls "barbarism" and indigeneity.

The grid was applied to the colonial towns of the East Coast of today's United States, as well, but there it followed different cultural patterns. The US-American urban planner Francis Violich highlights the differences: "While in the United States our cities embarked upon expansion without an established nucleus, without a sense of civic form or dignity inherited from history, most Latin American cities had such a pattern for future growth" (qtd. in Almandoz 2002: 1). Richard Sennett expands this argument in *The Conscience of the Eye: The Design and Social Life of Cities* by arguing that the US-American planners re-semanticized the grid and used it for a different purpose in neglect of the diversity of a given place:

> The grid can be understood, in these terms, as a weapon to be used against environmental character—beginning with the character of geography. In cities like Chicago the grids were laid over irregular terrain; the rectangular blocks obliterated the natural environment, spreading out relentlessly no matter the hills, rivers, or forest knolls that stood in the way. (1991: 52)

Although one can observe different uses of the grid in the Americas, coloniality formed their shared background. In Spanish America, the central idea behind the grid was to order the colonized space by rewriting the city in colonial—and often religious—terms and by creating visible power centers. In contrast to this concept, the English settlers needed a model that neutralized space in order to dominate it and its populations. In the process of the British colonies' (and later the United States') westward expansion, "the grid served as a weapon" (Sennett 1991: 76). Thus, in both cases the grid functioned to colonize space and install white supremacy. In Spanish America, this drive was based on a religious-monarchic social structure that aimed at centralizing power, state, and symbolic expression. In the US-American settler societies, the grid had the power to neutralize space and its ecological and ethnic diversity in order to render them controllable for an emerging capitalist economy.

EthniCity and Nation-Building

In Latin America, the urban forms and their cultural expressions changed in the context of the wave of colonial independence from Spain and Portugal between 1810 and 1825. While the urban form remained untouched in the first decades after Spanish rule, a new debate on urban planning emerged during the second half of the nineteenth century. These discussions took place against the backdrop of accelerated (economic) transnationalization processes and a growing integration of the Latin American societies into the world market, a process that was driven by new liberal elites inspired by the ideas of universal progress and free trade. In this context, the *damero* was conceived of as a colonial legacy that had to be overcome in the processes of modernization, progress, and political independence.

When searching for a new cultural paradigm appropriate for their nation-building projects, the white urban postcolonial elites of Latin America found a model and "muse" in contemporary France, especially Paris: "For the Frenchified elites of these cities, the invocation of Second-Empire Paris was thereby supposed to make possible their magic transformation from post-colonial city into real metropolis" (Almandoz 2002: 4). The urban imaginaries of Latin American white *criollo* elites were especially inspired by the modernizing urban development program of Baron Georges-Eugène de Haussmann, the prefect of the Seine *département* during the Second Empire. Haussmann modernized Paris by introducing monumental rectilinear streets and enforcing a socio-spatial restructuring of the city that sought to gain better political control of the urban population by segregating them according to function and class as well as sanitary and infrastructural criteria (Scott 1998, Harvey 2003).

Diverse cities ranging from Buenos Aires and Santiago de Chile to Mexico City were each imagined by their local elites as a sort of "Little Paris." In the case of one of the most influential restructuring projects *à la* Haussmann, the *Paseo de la Reforma* in Mexico City, the urban renewal of this avenue—originally built under the French emperor Maximilian—was related to a national spectacle, the 1910 centennial celebration of Mexico's War of Independence, as well as Mexico's application for hosting a World Exposition. The *Paseo de la Reforma* in Mexico City is said to be the first "copy" of the Parisian Boulevards in the Americas.

This urban modernization project was an attempt to inscribe Mexican history—represented in several monuments to famous personalities along the avenue—into the "master narrative" of Western universal progress and modernity, thus serving the identity political purpose of whitening Mexican culture under the dictatorship of Porfirio Díaz (Tenorio 1996). Haussmann's urban restructuring program addressed several issues that were also of key importance for the Western Hemisphere. In terms of functional aspects, one could promote urban renewal projects as steps toward the modernization of social structures which would prepare the former colonial cities for the industrial era. In this respect, urban renewal also provided the possibility for demographic restructuring and enlargement of the city. Beyond these functional as well as hygienic and technical aspects of urban planning, it is the specific narration of the city that explains the influential imaginary which Haussmann's Paris created. With its neo-classical architecture, neo-baroque plan, and connotation of modernity, it provided a model for articulating the Latin American elites' determined desire for nation-building through architecture and urban development.

These restructurings of nineteenth-century Latin American cities in the mold of Paris can be understood in imaginative terms as early expressions of theming, as a whole overarching motif; here Western European modernity—manifest in neo-classical architecture and neo-baroque structure—is used to bring about transformations of urban space, which aim to draw peripheral cities into a global net of cosmopolitan metropolises. In the Latin American case, it is a paradox that the decolonization from Spanish rule led to a voluntary cultural neocolonialism promoted by the white postcolonial elites, who eagerly adopted French patterns in urbanism. In this way, the coloniality of power, especially the classification and stratification of societies along lines of ethnic difference, remained untouched. Whitening and Western civilization continued to be the guiding principles, which were now understood in terms of a universal progress.

In the United States, it is especially the new, representative capital of Washington, DC, planned under the leadership of the French engineer Pierre Charles L'Enfant in 1791–92, that followed Western European "modernist" models (Reps 1965: 240–62). The L'Enfant plan outlined the city on the basis of a rectilinear grid interspersed with diagonally running avenues and representative squares and circles at major intersections. Even though L'Enfant's plan was quite in line with late eighteenth-century European architecture and city planning, the influence of European programs of urban modernization remained rather limited in North American urban development.

However, what did change in the United States, especially in the major cities in the East, was the emergence of a new dynamics in urban life in the course of the nineteenth century, as a large influx of European immigrants as well as freed or escaped slaves from the South led to accelerated urban growth. It is in this period that the typical ethnically-segregated structure of the US-American city was shaped, coined by the urban cultural politics of institutions and organizations as well as the daily practices of the urban newcomers. Chicago was soon considered an ideal-typical US-American city characterized by capitalist land use and ethnic segregation, both rendered paradigmatic in academic discourses by the Chicago School of Sociology (Plummer 1997). The grid continued to be the main settling pattern in the city, but, beginning in the 1870s, this model was extended not only in a horizontal but also in a vertical dimension. The idea of the dominance of a theoretically interminable space was now projected into the sky. At the beginning of the twentieth century, the cities of New York and Chicago in particular entered into a competition for the world's tallest building which would culminate in the completion of the Chrysler Building in 1930 and the Empire State Building in 1931, the world's tallest building for forty years. The skyscraper city represented a symbol for the nation's progress and modernity, thus representing the US-American way of life and claim to global economic and political leadership.

In the context of the attempts at import substitution and industrialization following World War II, Latin America witnessed a dynamics of modernization of its cityscapes modeled after US-American cities that went (as in the United States) hand in hand with a certain neglect of its historic city centers. The planning and construction of the new capital of Brazil, Brasília, in the midst of the unexplored hinterland represents the most radical project of modern utopian city planning in Latin America. Brasília provides a classic example of a centralized, state-run urban cultural politics that was carried out in close collaboration with experts from the field of cultural production, namely the architects Oscar Niemeyer and Lucio Costa, both influenced by the modernist paradigm of functional city planning as postulated by Le Corbusier. Brasília combines the modern ideal of technical planning and the monumental representation of a national utopia of modernity with a social reformist plan for housing (Holston 1989, Scott 1998). When seen from above, the central part of the city resembles an airplane, structured by a monumental axis of public buildings lined up along the main avenue, with housing units (*superquadras*) as wings and the government district forming the cockpit. This image suggests an intertextual reference to the societal development model proposed by W.W. Rostow, which compares social modernization with the take-off of an airplane.

The idea of central planning soon faded away when *favelas* emerged on the outskirts of Brasília's *plan piloto*, introducing "chaotic" and less functional urban cultural politics from below. There is, moreover, a certain kind of irony inherent in the fact that this iconic expression of the belief in state-led modernization and progress is now considered to be itself a nostalgic remnant of the past. Brasília holds the distinction of having been admitted to the UNESCO World Heritage List faster than any other place in the world, a mere 27 years after its completion in 1960. It thus represents the only city in the world built in the twentieth century to be awarded the status of Historical and Cultural Heritage of Humanity within the same century.

After World War II, many cities in the Americas lost their attraction as an integral space of daily life, leading to dynamics of suburbanization and urban sprawl. New (sub)urban forms of private and public life emerged, such as shopping centers and gated communities, which attempted not only to satisfy material needs, but also to create imaginative spaces. New "technopoles" (Castells and Hall 1994) of the post-Fordist information economy—like Silicon Valley—arose on the outskirts of cities. In fact, the idea of a single, integrated city had vanished and been substituted by a city that is fragmented in multiple ways. In order to describe the new urban geography, some scholars rely on island metaphors. Susan Christopherson understands the privatized places as "highly manipulated islands in the midst of anarchistic non-space" (1994: 410) and Peter Marcuse speaks of "insular spatial formations" (1997: 249). In contrast to the earlier grid model, gaining control over space by means of the layout of a city is no longer possible. The traditional territorial-geographical notion of space is substituted by a mobility-centered conception that attaches greater importance to spatial trajectories and the modes and time to cross space.

Insofar, I argue that the notion of isolation and enclosure in a "spatially concentrated area," where privileged groups "congregate as a means of protecting or enhancing" is misleading in the present case. On the contrary, from the perspective of middle- and upper-class consumers, places such as gated communities, malls, theme parks, airports, and the like are not isolated, but linked to one another functionally by highways and imaginatively through the production of a themed landscape. Using models of perception, one can understand the iconic spatial elements of neo-monumental postmodernist architecture as serving as symbolic landmarks which endow the landscape with new meaning. Therefore, I propose to speak of an urban landscape in the form of a fractal archipelago, whose form depends on the identitarian position of the observer.

Thus, it is certainly true, as Néstor García Canclini (1995) states for the case of Mexico City, that the overall sense of the city is fragmented and that urban spectacles cannot easily reconstitute the city's lost wholeness. Nevertheless, there are various notions of the city that depend on socioeconomic and cultural milieus, thus producing new grids which do not control space form a central perspective, but appropriate it along (high)ways and communication flows in multi-perspective forms. With this fragmentation of the imaginary of a homogenous city which has the aim to "civilize" and nationalize different ethnicities, a new space opens up for articulating different identitarian forms of expression in urban spaces.

Consuming EthniCity on the Global Grid

One of the most important dynamics that shapes contemporary societies, and thus urban spaces, is undoubtedly the accelerated globalization process. On the one hand, Los Angeles has been seen as a model for a dispersed post-metropolis (Davis 2006, Dear and Schockman 1996) that is—following Edward Soja (1989)—characterized by restructuring processes, including the emergence of a postindustrial economy, globalized cosmopolitan cultural patterns, a "carceral" architecture revealing social exclusion processes, urban deterritorialization (sim city), and ethnic fragmentation (fractal city). On the other hand, one can identify a new concentration of goods, capital, people, and information in the cities that have come to be nodal points in a global network economy. A dis-embedding of these cites from the framework of the nation state can be observed; instead they are re-embedded in a global archipelago.

Thus a new form of the grid has emerged on a global scale that is constituted by the flows of information, goods, capitals, and people, with cities functioning as its major points of articulation. This grid is narrower in Northern, "developed" societies, while many areas of the global South remain in a peripheral position or are even excluded. Saskia Sassen defines global cities as "cities that are strategic sites in the global economy because of their concentration of command functions and high-level producer service firms oriented to world markets; more generally cities with high levels of internationalization in their economy and in their broader social structure" (1991b). Nevertheless, besides this strict definition of global cities as the commanding center of world economy (Sassen 1991a, 1991b), one can observe a general dynamic of cities positioning themselves on the global grid in order to become global cities on various levels of the transnational urban hierarchy. Some cities aim to establish themselves as cultural capitals that appeal to new cosmopolitan elites. According to Louise Johnson, one can understand a "cultural capital" as "a city which has recently and consciously made the arts (and often related Cultural Industries) central to its society, economy, urban form and place identity" (2009: 6).

This renewed attention to the city produces an urban dynamic that I would like to call "return to the center" (Herzog 2006, Rojas 2004). "Return to the center" entails the reoccupation of spaces in centrally located (downtown) urban districts by hegemonic groups, often through processes of gentrification, urban renewal, slum clearance, and the establishment of new urban consumer lifestyles and identities in the centers. Obviously the idea of this recent return to the center is perceived from the perspective of the hegemonic middle and upper classes, as the center in fact was never abandoned. Instead it can be argued that in the midst of the twentieth century, one could witness a "move to the center" by subaltern, often indigenous peasants migrating from rural areas in Latin America while in North America the city centers were populated by lower-class city-dwellers, especially from ethnic/racial minorities. Now, middle-class and creative milieus are returning, and the subaltern social strata are being increasingly pushed aside.

In a study conducted for the Inter-American Development Bank, Eduardo Rojas and his collaborators Eduardo Rodríguez and Emiel Wegelin (2004) discuss theoretical approaches, instruments, and case studies of processes and projects that aim at urban renewal in central districts of different cities. From their international perspective, they discuss three categories of inner-city revitalization, all of which relate to the characteristics of the areas in question. First, they identify programs targeting relatively centrally located districts that had been previously abandoned, mainly by processes of de-industrialization. Examples of this include Puerto Madero in Buenos Aires, the Elbe Zone in Hamburg, and the Docklands in London.

The second category consists of programs designed to modify the dynamics of the city in order to achieve balanced urban growth. These programs seek to stimulate the renovation of deteriorated neighborhoods close to city centers. The third category includes programs that restore central (downtown) city areas. These interventions are particularly large and complex and include the coordination of different programs. Here the report examines the cases of the Penn Quarter in Washington D.C., eastern Santiago de Chile, and the historic city center of Quito. As the authors of the study note, this third category is of special importance for urban processes in Latin America, because nearly all Latin American countries face the problems of urban decay in city centers (Rojas 2004: 130).

Needless to say, this revival of urban centers in a post-Fordist global cultural economy is a highly conflictive process, as it is related to an appropriation of space hitherto largely occupied by subaltern, often impoverished social strata. Since the 1980s, gentrification has been the word of choice to denote this process of re-appropriation of inner-city districts and near-downtown neighborhoods by the middle and upper classes, a process that profoundly influences the ethnic cityscape. Neil Smith argues that these are elements of a "revanchist city" (1996) that can be understood as a reversal of the earlier "white flight," now moving from a culturally barren suburbia to the culturally fertile urban centers. Nevertheless, this process does not lead to a straightforward annihilation of ethnic difference and "whitening" of the cities; in interactions which are often more complex and convoluted, for some "ethnic" or "alternative" districts, cultural and ethnic difference from the mainstream serve as key factors of a neighborhood's attraction—desirable features which must be maintained in the process of gentrification.

It is remarkable that these urban transformations take place in the heydays of expansion of the field of identity politics in the Americas (Thies and Kaltmeier 2009). Multiculturalism and struggles for recognition of ethnic difference profoundly inform the cultural political dynamics that have rendered the formerly unmarked ethnic category of whiteness visible as one ethnicity among others (Bonnett 2000). Furthermore, one can observe increasing flows of transnational migration into the urban centers which have led to the emergence and ethnic diasporas in multiple urban ethnic spaces—spaces which can be called, following Arjun Appadurai (1996), global ethnoscapes. While the modern city was structured according to the primacy of industrial production, in the postmodern city consumption is the guiding paradigm, provoking a proliferation of cultural economies that especially concerns urban areas. A large body of critical scholarship focuses on the changing and expanding character of cultural industries in post-Fordist urban formations, ranging from economic political studies to socio-culturally orientated approaches that address the material production of literature and media (Rifkin 2000, Sennett 2006, Mato 2007).

This body of work articulates a broader dynamics in post-Fordist societies, by which Cultural Studies scholars and cultural sociologists argue for the existence of an ongoing de-differentiation between economy and culture in late capitalist societies (Short and Kim 1999). George Yúdice speaks of an "expediency of culture" (2003) to capture the growing strategic use of culture for economic and political purposes. Moreover, culture is not only used strategically, but it comes to be epistemic in the sense that many other conflicts and negotiations are interpreted through a cultural frame. In a similar vein, Daniel Mato (2007) argues that the classical definition of culture industry, introduced by Theodor W. Adorno and Max Horkheimer (1947), is now being substituted by a dynamics in which all industries make use of culture and thus to a certain extent become cultural industries of their own.

These arguments can be applied to the city, especially in a context in which the city is understood as a kind of business or company with marketing departments. Cities, regions, and institutions related to them increasingly use identitarian political strategies such as "imageneering," "branding," and "theming" to position themselves in the global competition for best (business) location. Their objective is to increase the attraction of their respective locations—their cities—for investors, corporate employers, and tourists by creating a specific "sense of place" (Bryman 2004, Gottdiener 1998, Zukin 1995, Gold and Gold 2005) as a "unique selling point" for their cities in the global cultural market.

Sociological approaches affiliated with regulation theory connect these transformations in urban spaces to an ongoing process of transition from the regulatory mode of Fordism to post-Fordism. In the context of cities in the Americas, the shift from the primacy of production to that of consumption is of particular importance (Bauman 2007). In consumer societies (Bauman 2007, Baudrillard 1998, Featherstone 1991, García Canclini 2001), consumption is only perceived secondarily as a means to satisfy material needs. Instead, the consumption of goods can be seen as an indicator of identitarian belonging to certain social groups and/or distinction from others, based on the often intersecting factors of social class status, ethnicity, gender, age, and the like. Some scholars underline the integrationist function of consumption, as it is postulated in the manifesto of consumption presented by Norbert Bolz, who hints at the pacifying and conflict-solving role of consumption as a civilizing achievement of capitalism (2002). In sharp contrast to this and other social-utopian imaginaries of consumption-based mechanisms for social integration, the sociologist Zygmunt Bauman notices a "postmodern privatization of identity problems" (2007) that leads to a social imperative to consume as the privileged means to identity formation.

These general patterns of postmodern consumer society are closely related to identity politics. The imperative to consume goes hand in hand with the social imperative of identitarian "choice" (Hollinger 1995) and the shaping and self-marketing of identities. While the liberal plea for choice in a post-multicultural society places the autonomous individual at the center, Bauman argues that this individual could be better conceived of in terms of processes of subjectivation which no longer obey the disciplining patterns of the Fordist factory regime, but instead follow the (self-)control of consumer society. Here, identity itself comes to be a consumable commodity. Those who lack (economic or cultural) access to consumption are excluded from constructing socially accepted identities, an exclusion unknown in its extent in production-based societies.

In nearly all urban areas, one can observe the cosmopolitan strategy to associate commodities with the world's metropolitan centers of cultural production by making their metropolitan origin a significant part of their consumption value (Hannerz 1996). This linkage is achieved by valorizing particular brand names such as Starbucks, Hard Rock Cafe, or Nike, to name but a few, associated with new cosmopolitan lifestyle identities, and it often goes hand in hand with the tendency to identify or create unique consumer items connected to a specific city through which a public image of that city can be constructed.

These developments are further related to a transformation in the understanding of processes of production and consumption. While Fordism placed standardized products for a large number of consumers on the market, post-Fordist societies are marked by the fragmentation of the market into distinct sectors, each of which appeals to specific, numerically smaller consumer identities (Holland and Gentry 1999, Ogden, Ogden, and Schau 2004). As Jan Lin points out in his study on Chinatowns, "the function of ethnic enclaves in the global city" should include recognizing "the role that ethnic enterprises have played in reviving industry, warehousing, and retailing districts of the central city" (1998: 314). In Chicago, for example, increasing Chinese-based investments reflect the significance of a global ethnic economic network in the diaspora now commonly labeled "overseas Chinese." This network relies to a great extent on the post-Fordist marketing and exoticization of ethnic difference, in which ethnicity is used as a resource to sell supposedly "authentic" places and goods such as ethnic cuisine (Narayan 1997, hooks 1992). The revalorization of ethnic places such as Chinatowns implicates them in broader strategies of transnational accumulation employed by city marketing departments working to project a multicultural image of their city that fits into the preferences of a post-national, cosmopolitan elite. This phenomenon is tightly linked to the growing economic impact of city tourism, for which "authentic" urban cultures and ethnic neighborhoods—especially those in historic city centers—are considered key factors (Meethan 2001, Gotham 2007, Lash and Urry 1994, Urry 1990, 1992).

Although the growing tolerance toward and acceptance of different ethnic groups and their cultural expressions in the city contributes to a democratization of urban life, it is important to mention the limits of these multicultural urban politics mentioned above. In post-9/11 United States cities, Chinese clothing, architectural ornaments, ethnic food, or Latino carnivals may be seen as an enrichment of a city's cosmopolitan diversity, whereas a bearded "Oriental" sporting a turban would cause fear and show the limits of tolerance. As Loïc Wacquant points out with regard to the urban outcast in "hyper-ghettoes" (2007), the intersection of class and ethnicity as well as the processes of fragmentation and postmodern anomy pose further limits to multiethnic tolerance. In short, one can speak of a "permitted multiculturalism" or a "permitted ethnicity" that is contingent on being consumable. In the postmodern urban cultural economy, ethnicity must sell in order to gain the right of recognition.

The purpose of the present volume, *Selling EthniCity: Urban Cultural Politics in the Americas*, is to explore the importance of ethnicity and the cultural economy in the post-Fordist city of the Americas. Cultural, political, and economic elites make use of cultural and ethnic elements in city planning and architecture in order to construct a unique image for a particular city; they create urban festivals and spectacles, promote international cultural heritage programs, and involve cultural producers in performances and visualizations of a given city in order to attract (trans)national tourists, investors, and upscale residents. These developments are recoded in daily life and contested by the cultural politics of urban movements.

The present volume approaches this increasing importance of culture in the city in four ways: First, within the growing entertainment industry, art and museum scene, tourist sector, and creative economy, one can observe the increasing importance of cultural industries for the urban economy. Secondly, one has to point out the importance of urban spectacles that finds its expression in festivals, fairs, and exhibitions as well as in sports, music, and other cultural events. Third, one can note a rapid emergence of (new) aesthetic, cultural, and symbolic landscapes in the city. These include processes of theming, imagineering, and urban renewal and revitalization, especially in the context of cultural heritage. A fourth aspect is related to culture as a governmental strategy, which refers to a wide array of practices from *latte-macchiatoization* via surveillance to gentrification and the forced expulsion of "dangerous populations."

As an overly complex space of interaction and signification, the contemporary city cannot be conceptualized from only one academic discipline; therefore, an interdisciplinary dialogue and transdisciplinary research perspective are of particular importance when exploring the intersection of urban development and cultural politics. For this purpose, *Selling EthniCity* assembles contributions from various academic disciplines such as sociology, social anthropology, human geography, history, urban studies, architecture, cultural studies, media studies, and literary studies in order to contribute to the new research paradigm of Inter-American Studies as transnational area studies.

Works Cited

Almandoz, A. 2002. *Planning Latin America's Capital Cities. 1850–1950*. London: Routledge.

Alvarez, S., Dagnino, E. and Escobar, A. (eds.) 1998. *Cultures of Politics, Politics of Cultures: Re-Visioning Latin American Social Movements*. Boulder: Westview Press.

Appadurai, A. 1996. *Modernity at Large: Cultural Dimensions of Globalization*. Minneapolis: University of Minnesota Press.

Baudrillard, J. 1998. *The Consumer Society: Myths and Structures*. London: Sage.

Bauman, Z. 2007. *Consuming Life*. Cambridge: Polity Press.

Benevolo, L. 1978. *The Architecture of the Renaissance*. London: Routledge and Kegan Paul.

Bolz, N. 2002. *Das konsumistische Manifest*. München: Fink.

Bonnett, A. 2000. *White Identities: Historical and International Perspectives*. Harlow: Prentice Hall.

Bryman, A. 2004. *The Disneyization of Society*. London: Sage.

Castells, M. and Hall, P. 1994. *Technopoles of the World. The Making of 21st-Century Industrial Complexes*. London and New York: Routledge.

Christopherson, S. 1994. The Fortress City: Privatized Spaces, Consumer Citizenship, in *Post-Fordism: A Reader*, edited by A. Amin. Oxford: Blackwell, 409–27.

Davis, M. 2006. *City of Quartz*. Brooklyn, NY: Verso Books.

Dear, M., Schockman, E. and Hise, G. (eds.) 1996. *Rethinking Los Angeles*. Thousand Oaks, London, and New Delhi: Sage.

Eisenstadt, S.N. 2002. The First Multiple Modernities: Collective Identity, Public Spheres and Political Order in the Americas, in *Globality and Multiple Modernities*, edited by L. Roniger and C. Waisman. Brighton: Sussex Academic Press, 7–28.

Featherstone, M. 1991. *Consumer Culture and Postmodernism*. London: Sage.

García Canclini, N. 1995. Mexico: Cultural Globalization in a Disintegrating City. *American Ethnologist*, 22(4), 743–55.

García Canclini, N. 2001. *Consumers and Citizens: Globalization and Multicultural Conflicts*. Minneapolis: University of Minnesota Press.

Gilroy, P. 1993. *The Black Atlantic: Modernity and Double Consciousness*. London: Verso.

Gold, J. and Steven, W. (eds.) 1994. *Place Promotion: The Use of Publicity and Public Relations to Sell Towns and Regions*. Chichester: John Wiley & Sons.

Gold, J. and Gold, M.M. 2005. *Cities of Culture: Staging International Festivals and the Urban Agenda, 1851–2000*. Burlington, VT: Ashgate.

Gotham, K. 2007. *Authentic New Orleans: Tourism, Culture, and Race in the Big Easy*. New York: New York University Press.

Gottdiener, M. 1998. *Postmodern Semiotics: Material Culture and the Forms of Postmodern Life*. Oxford: Blackwell.

Hannerz, U. 1996. *Transnational Connections: Culture, People, Places*. London and New York: Routledge.

Harvey, D. 2003. *Paris, Capital of Modernity*. London: Routledge.

Herzog, L. 2006. *Return to the Center: Culture, Public Space, and City Building in a Global Era*. Austin: University of Texas Press.

Holland, J. and Gentry, J. 1999. Ethnic Consumer Reaction to Targeted Marketing: A Theory of Intercultural Accommodation. *Journal of Advertising*, 28, 65–77.

Hollinger, D. 1995. *Postethnic America: Beyond Multiculturalism*. New York: Basic Books.

Holston, J. 1989. *The Modernist City: An Anthropological Critique of Brasilia*. Chicago: University of Chicago Press.

hooks, b. 1992. Eating the Other: Desire and Resistance, in *Black Looks: Race and Representation*, edited by B. Hooks. Boston: South End Press.

Horkheimer, M. and Adorno, T. 1947. *Dialektik der Aufklärung: Philosophische Fragmente*. Amsterdam: Querido.

Johnson, L.C. 2009. *Cultural Capitals: Revaluing the Arts, Remaking Urban Spaces*. Farnham and Burlington, VT: Ashgate.

Kaltmeier, O., Kastner, J., and Tuider, E. (eds.) 2004. *Neoliberalismus—Autonomie—Widerstand: Analysen Sozialer Bewegungen in Lateinamerika*. Münster: Westfälisches Dampfboot.

Kearns, G. and Philo, C. (eds.) 1993. *Selling Places: The City as Cultural Capital, Past and Present*. Oxford: Pergamon Press.

Kymlicka, W. 1995. *Multicultural Citizenship*. Oxford: Oxford University Press.

Lash, S. and Urry, J. 1994. *Economies of Signs and Space*. London, Thousand Oaks, and New Delhi: Sage.

Lin, J. 1998. Globalization and the Revalorizing of Ethnic Places in Immigration Gateway Cities. *Urban Affairs Review*, 34(2), 313–39.

Mallon, F. 1992. Indian Communities, Political Cultures, and the State in Latin America, 1780–1990. *Journal of Latin American Studies*, 24, Quincentenary Suppl., 35–53.

Marcuse, P. 1997. The Enclave, the Citadel, and the Ghetto: What has changed in the Post-Fordist US city. *Urban Affairs Review*, 33(2), 228–64.

Mato, D. 2007. Todas las industrias son culturales: crítica de la idea de "industrias culturales" y nuevas posibilidades de investigación. *Nueva época*, 8, 131–53.

Meethan, K. 2001. *Tourism in Global Society: Place, Culture, Consumption*. Basingstoke and New York: Palgrave.

Mignolo, W. 2000. *Local Histories / Global Designs*. Princeton: Princeton University Press.

Mignolo, W. 2005. *The Idea of Latin America*. London: Blackwell.

Narayan, U. 1997. *Eating Cultures: Incorporation, Identity, and Indian Food*. New York and London: Routledge.

Ogden, D.T., Ogden J.R., and Schau H.J. 2004. Exploring the Impact of Culture and Acculturation on Consumer Purchase Decisions: Toward a Microcultural Perspective. *Academy of Marketing Science Review*, 3.

Ostendorf, B. (ed.) 2002. *Transnational America: The Fading of Borders in the Western Hemisphere*. Heidelberg: Winter.

Plummer, K. (ed.) 1997. *The Chicago School*. London: Routledge.

Quijano, A. 2000. Colonialidad del poder, eurocentrismo y América Latina, in *Colonialidad del saber, eurocentrismo y ciencias sociales*, edited by E. Lander. Buenos Aires: CLACSO, 342–86.

Rama Á. 1984. *La ciudad letrada*. Hanover: Ed. Del Norte.

Reps, J.W. 1965. *The Making of Urban America: A History of City Planning in the United States*. Princeton: Princeton University Press.

Rifkin, J. 2000. *The Age Of Access: The New Culture of Hypercapitalism, Where All of Life is a Paid-For Experience*. New York: Tarcher.

Rojas, E. 2004. *Volver al centro: la recuperación de áreas urbanas centrales*. Washington, DC: Inter-American Development Bank.

Sadowski-Smith, C. and Fox, C.F. 2004. Theorizing the Hemisphere: Inter-Americas work at the Intersection of American, Canadian, and Latin American Studies. *Comparative American Studies*, 2(1), 41–74.

Sassen, S. 1991a. *The Global City: New York, London, Tokyo*. Princeton: Princeton University Press.

Sassen, S. 1991b. *The Global City: Strategic Site/New Frontier* [Online]. Available at: http://www.india-seminar.com/2001/503/503%20saskia%20sassen.htm [accessed: July 12, 2010].

Schafhausen, N. 2010. Unsere frivole Begierde nach Täuschung. *Süddeutsche Zeitung*, February 24, 12.

Scott, J. 1998. *Seeing Like a State: How Certain Schemes to Improve the Human Condition Have Failed*. New Haven: Yale University Press.

Sennett, R. 1991. *The Conscience of the Eye: The Design and Social Life of Cities*. New York: Faber and Faber.

Sennett, R. 2006. *The Culture of the New Capitalism*. New Haven: Yale University Press.

Short, J.R. and Kim, Y. 1999. *Globalization and the City*. Addison: Harlow.

Smith, N. 1996. T*he New Urban Frontier: Gentrification and the Revanchist City*. London and New York: Routledge.

Soja, E. 1989. *Postmodern Geographies: The Reassertion of Space in Critical Social Theory*. Brooklyn, NY: Verso.

Tenorio Trillo, M. 1996. 1910 Mexico City: Space and Nation in the City of the *Centenario. Journal of Latin American Studies*, 28, 75–104.

Thies, S. and Kaltmeier, O. 2009. From the Flap of a Nutterfly's Wing in Brazil to a Tornado in Texas? Approaching the Field of Identity Politics and its Fractal topography, in *E Pluribus Unum? National and Transnational Identities in the Americas / Identidades nacionales y transnacionales en las Américas*, edited by S. Thies and J. Raab. Münster and Tempe, AZ: Lit Verlag and Bilingual Review Press, 25–46.

Thies, S. and Raab, J. (eds.) 2009. *E Pluribus Unum?: National and Transnational Identities in the Americas / Identidades nacionales y transnacionales en las Américas*. Münster and Tempe, AZ: Lit Verlag and Bilingual Review Press.

Urry, J. 1990. *The Tourist Gaze: Leisure and Travel in Contemporary Societies*. London: Sage.

Urry, J. 1992. The Tourist Gaze Revisited. *American Behavioral Scientist*, 36(2), 172–86.

Wacquant, L. 2007. *Urban Outcasts: A Comparative Sociology of Advanced Marginality*. Cambridge, UK: Polity Press.

Yúdice, G. 2003. *The Expediency of Culture*. Durham, NC: Duke University Press.

Zukin, S. 1995. *The Cultures of Cities*. Oxford and Malden, MA: Blackwell.

PART I
The Spectacular City and the
Performance of Ethnicity

Introduction to Part 1

The Spectacular City and the Performance of Ethnicity

Olaf Kaltmeier

Olympic Games, World Cups, and World Fairs, as well as theater, music, and film festivals are major events that represent what has been called the "festivalization of urban politics" (Häußermann and Siebel 1993, Gold and Gold 2005). What characterizes these urban cultural events is a thematic focus of urban politics on spatially and temporally limited projects. Spectacular, future-oriented buildings and large-scale urban development programs in the host cities often accompany these internationally-oriented projects, such as expositions and major sporting events, which momentarily grace a certain location and then travel between cities around the globe. In this respect, such events or major projects are to be regarded, *inter alia*, as instruments of urban development policy.

While large spectacles have existed since ancient Europe or amongst Pre-Columbian Andean cultures, the above-mentioned spectacles, which form part of a post-Fordist urban economy, differ from their predecessors in many ways. The post-Fordist spectacles are subject to a "process of commodification, homogenization and rationalization of time and space" (Gotham 2005: 234), and their overall goal is to make profit. This profit includes both short-term revenues from the spectacle (advertising revenues, donations, ticket sales) as well as project funds acquired from public and private institutions, and medium-term public image gains that are to be capitalized upon by attracting (cultural) tourism as well as (creative) industries. In order to stand their ground in the competitive market of (international) media attention and to distinguish themselves from the plenitude of festivals organized today, cities strategically employ ethnicity as a unique selling point, as could be witnessed most recently in the 2010 Olympic Winter Games which the city of Vancouver organized in cooperation with the First Nations at home in the region.

Neo-Marxist thinkers have been criticizing the spectacularization of cities for a long time (Zukin 1995, Debord 1995). Foremost among these is Guy Debord's visionary essay on the "society of the spectacle," which criticizes the fetishization and alienation in a society that increasingly articulates itself through spectacles. According to Debord, the "spectacle" marks the very moment in which commodities engross all social life and one can no longer see beyond the world of commodities (1995: § 42). However, as Kevin Fox Gotham in particular has shown, spectacles also open up opportunities for resistance as well. Taking the New Orleans Mardi Gras as an example, Gotham describes the extent to which "local actors can use urban spectacles for positive and progressive ends, including launching a radical critique that exposes the deprivations of class and racial inequality" (2005: 235). Debord himself has discussed the possibilities of critique inherent in the strategy of *detournement*. Here, the originally intended hegemonic meanings of images are placed in a new context in order to confront images of abundance with a reality of impoverishment and thus better reveal the oppressive character of consumer capitalism (1995: 206–11). Cultural historian Celeste Olalquiaga (1992: 82–6) provides the telling example of a resistance-oriented contribution of a group of cultural producers to the 1987 Samba Festival in Rio de Janeiro. The group constructed a futuristic city called "Tupinicópolis," which, as Anton Rosenthal puts it, "satirized contemporary high-tech culture, postindustrial progress, Hollywood, and consumerism, all the while maintaining a self-consciousness of being a spectacle and adopting an aesthetic that upended First World perceptions of Latin America" (2000: 62). This example poignantly represents the strategy of *detournement* as proposed by Debord, in which the organization of a spectacle according to the rules of post-Fordist consumer society is used to question common stereotypes about Latin America and its indigenous populations.

In *The Spectacular City*, social anthropologist Daniel Goldstein works out a topical understanding of the spectacle that distinguishes itself from the concept of spectacles as commodities and instead emphasizes the inversion of social order:

> Spectacles, like other public events, are systems for not only the performance but also the creation or transformation of social order ... by calling explicit attention to things or relationships that often escape notice in the flow of everyday life, they can also operate as a critique of the existing social system by presenting alternative forms of living and social ordering. (2004: 16)

Thus, in many cities one can also find non-commodifiable and easily organizable spectacles, often realized by urban popular cultures or social movements. In Latin American cities in particular, with their frequently employed marked grid of streets and avenues, the plaza (square) has a central place in the urban landscape and serves as a favorite locus of urban spectacles (Rosenthal 2000). However, as Goldstein argues (2004) when drawing upon the example of the Bolivian city of Cochabamba, these spectacles include not only carnivals, parades, demonstrations, and processions but also such violent forms as public lynching.

According to the cultural theorist George Yúdice (2003), since the middle of the twentieth century and beginning with the Black Civil Rights movement in the United States, there has been an ever-increasing connection linking identity politics and performativity which has become a "social imperative to perform" for the different ethnic (and other) groups pursuing this type of politics. The following chapters explore precisely this performative dimension of urban spectacles. In "Carnival Redux," John Gold analyzes the transformations of contemporary

US-American urban festivals, taking the 2006 Mardi Gras celebration in New Orleans—the first carnival after Hurricane Katrina had devastated the city—as his starting point. Gold draws upon ongoing research into the history of staging non-ambulatory cultural festivals, to which he adds more general reflections on the contemporary experience of the ways in which American cities have embraced an ancient yet continually changing festival form, before he finally returns to explore the lessons of the first post-Katrina New Orleans Mardi Gras celebration.

Wilfried Raussert and Christina Seeliger's chapter "What Did I Do to Be So Global and Blue?" explores the commodification of Blues music in the context of a growing urban tourism in Chicago. The text discusses the multiple and ambivalent patterns of a politics of authenticity. Though city marketing activities promote the Blues as part of a (commodified) African American heritage, on the one hand, become ethnically and culturally more diversified, while on the other hand, Blues music is now largely ignored by urban African American youth, who instead use Hip Hop to articulate their identity.

Jens Kastner discusses the political and economic uses of graffiti/street art as a cultural expression situated on the edge between art and daily life. In particular he explores the cultural appropriation of public spaces by urban social movements and cultural activists in the context of the resurrection in Oaxaca, Mexico.

Social anthropologist Juliana Ströbele-Gregor discusses the spectacular acts of violence committed in the Bolivian capital of Sucre on May 24, 2008 by white and mestizo citizens against indigenous campesinos who had traveled to the capital for the festivities commemorating the 1809 Revolution of Chuquisaca. The analysis of this eruption of violence is placed in the wider political context and configuration of social conflict in Bolivia with a particular emphasis on strategies of street politics, discourses of strategic essentialism, and symbolic demonstrations of power.

Works Cited

Debord, G. 1995 [1967]. *The Society of the Spectacle*, translated by Donald Nicholson-Smith. New York: Zone Books.

Gold, J. and Gold, M.M. 2005. *Cities of Culture: Staging International Festivals and the Urban Agenda, 1851–2000*. Burlington, VT: Ashgate.

Goldstein, D. 2004. *The Spectacular City: Violence and Performance in Urban Bolivia*. Durham, NC: Duke University Press.

Gotham, K. 2005. Theorizing Urban Spectacles: Festivals, Tourism and the Transformation of Urban Space. *City*, 9(2), 225–46.

Häußermann, H. and Siebel, W. (eds) 1993. *Festivalisierung der Stadtpolitik: Stadtpolitik durch große Projekte*. Opladen: Westdeutscher Verlag.

Olalquiaga, C. 1992. *Megalopolis: Contemporary Cultural Sensibilities*. Minneapolis: University of Minnesota Press.

Rosenthal, A. 2000. Spectacle, Fear, and Protest: A Guide to the History of Urban Public Space in Latin America. *Social Science History*, 24(1), 33–73.

Yúdice, G. 2003. *The Expediency of Culture*. Durham, NC: Duke University Press.

Zukin, S. 1995. *The Cultures of Cities*. Oxford and Malden, MA: Blackwell.

Chapter 1

Carnival Redux: Hurricane Katrina, Mardi Gras and Contemporary United States Experience of an Enduring Festival Form

John R. Gold

Mardi Gras revelers along St. Charles Avenue and Canal Street on Monday night whooped for the marching bands, hollered for celebrities such as Dan Aykroyd, applauded the lavish floats and cried out for the trinkets tossed by costumed riders as they had for decades. But behind all the merriment and the masks something was missing.

New Orleanians are tired. They are distracted. On the face of it, they seem as lighthearted as ever. But they are not. And so it is with Mardi Gras—the two-week pre-Lenten celebration that ends today, "Fat Tuesday." It is exuberant on the outside, strange and different and diminished by loss on the inside. (Weeks and Whoriskey: 2006)

The staging of Carnivals, as Natasha Barnes (2000: 93) observed, is changing as demands for enhanced spectacle and performance erode the earlier religious and symbolic meanings of the events. Barnes was specifically concerned with the exploitative display of female masqueraders by Trinidadian Carnival organizers, but it is argued more generally that the modern Carnival, whether in Trinidad (Grant 2004) or elsewhere (Twitchell 1992), is a shadow of its precursors from earlier ages. While there is nothing wholly new about such arguments (e.g., Anon. 1875), there remains a deep concern that pressures from commerce and the media have debased the deeper meanings of Carnival, blurred its significance for the host city, and corroded the bonds that connect its participants, in varying ways, to the broader social order.

This chapter, however, draws on ongoing research into the history of staging non-ambulatory cultural festivals (Gold and Gold 2012) to argue that the experience of contemporary US-American cities—far removed in space and time from Carnival's medieval European origins—shows it to be an enduring festival form that is sufficiently malleable to absorb change while retaining its defining transgressive characteristics. This chapter contains three main sections. The first provides context by tracing the origins and spread of Carnival, a festival with pre-Christian roots, from Catholic Europe to North America. The next section analyzes the reconfigurations of Carnival that have arisen through the increasingly multifarious role of such festivals in US-American urban life as new agendas have emerged. Nevertheless, it emphasizes the way that changes and innovations have been absorbed within a form that is still recognizable with that which has endured through the centuries. The final part places these thoughts into sharper perspective by considering the lessons that can be learned from an analysis of the 2006 New Orleans Mardi Gras celebrations, staged less than six months after Hurricane Katrina wrought devastation upon the city and its suburbs.

The Nature of Carnival

There is no particular controversy about the original meaning of "Carnival" as a festival within Medieval Catholic Europe, although the exact etymology of the word is disputed. Its essential roots are variously taken to be the Medieval Latin verb *carnelevare*, meaning "to take away [or to remove] meat," the Italian *carne lasciare*, "leaving or forsaking flesh," or a number of similar variants. Whichever is favored, it is clear that by the twelfth century the name was firmly associated with the eve of Ash Wednesday and relates to the injunction to Christians to fast and be penitent during the forty days of Lent. Yet Carnival, like the pre-Christian festivals which it was said to replace (Harris 2003: 7–9), was the antithesis of solemnity. Rather, it came to mean the pre-Lent festival that gave an opportunity for indulgence and feasting. Carnival observances featured the world turned upside down, with playful inversions of the social order (e.g., Scribner 1978). The pauper was crowned king, women dressed as men and vice versa, and people commonly wore masks to create anonymity (actual or symbolic). Role play abounded as part of an order that all knew was temporary. Municipal regimes and church authorities did not necessarily like the transgression of social norms and the connotations of permitted misrule that became associated with the Shrovetide festival (Humphrey 2001) but were prepared to permit it as a lesser evil in order to enforce the asceticism of Lent and the sanctity of Easter.

The festivities were widely recorded in European art and literature. Paintings by Pieter Brueghel, Jan Miel, Hieronymus Bosch, and others illustrate the frenzy and exhilaration of festive observances, although their depictions tend to carry heavy overtones of moral judgment about the attendant excess (Yates 1997: 65). The Russian philosopher Mikhail Bakhtin (1984) locates somewhat different interpretations in the detailed, if often scatological descriptions of Renaissance carnivals found in the writings of the sixteenth-century French writer François Rabelais. Rather than viewing Carnival principally through the lens of transient disorder and debauchery, Bakhtin argues that it might instead be read as an expression of integration and even equality. In his words: "Carnival is the place for working out a new mode of interrelationship between individuals. … People who in life are separated by impenetrable hierarchical barriers enter into free and familiar contact on the carnival square" (1984: 123). Carnival, then, was a place-based, transgressive public festivity that was the outward expression of the workings of a social order. However anarchic it might look, it was actually firmly controlled by custom and ritual (see also Ravenscroft and Gilchrist 2009).

The notion that Carnival symbolism and ritual might be held up as constituting a "royal road" to developing an understanding of cultural "deep structure" (Gilmore 1998: 3) is shared by a variety of anthropological (Kertzer 1988, Mintz 1997) and historical writers (Ladurie 1980, Stallybrass and White 1986, Hennelly 2002). Although it might be argued that this construction overemphasizes the social purposes of carnival as against its ludic and subversive elements, there is no doubt that the European usage of the word "Carnival" was inescapably linked to this tradition. To some extent, of course, it can be argued that the European tradition itself has loosened in this regard, particularly with respect to both the timing and content of Carnival. For instance, by the eighteenth century the Venetian *carnevale*, including the wearing of masks, continued for around six months of the year as an adjunct to attracting the growing numbers of Grand Tour visitors that traveled to Italy intent on absorbing the sights and pursuing carnal delights (Tanner 1992: 43). In its different way London's Notting Hill Carnival, established in 1966 and now Europe's biggest street festival, is held each August on Bank Holiday rather than at the start of Lent, with practices imported from the West Indies as an assertion of minority identity (Jackson 1988). Nevertheless, Carnival overwhelmingly remains a time of feasting and merriment, with the populace briefly able to take advantage of socially accepted customs to cross the boundaries of what the dominant group in society ordinarily permits or would regard as good taste.

By contrast, the usage of the term "Carnival" in American parlance is traditionally less restrictive. It is certainly the case that was first applied to the staging of Mardi Gras festivities along the Gulf Coast of French North America, with the first festivals established in Mobile, Alabama, in 1703 and in New Orleans in 1718. The term, however, soon lost any sense of exclusivity. At its loosest, it became an alternative word to describe traveling fairs or exuberant festivities. Elsewhere, it has taken on more specialized meanings. In James Twitchell's *Carnival Culture* (1992), for instance, "carnival" is treated as an analogy for the operations of the media—notably publishing, motion pictures, and television—through which images are shown and for the industries that transport these images, the audience that makes up the traffic, and the critics who comment on the process. In a very different way, Philip McGowan sees "American carnival" as an adjunct of a racialized gaze, "an interpretative and representative" way of seeing "activated at both the conscious and subconscious levels to consolidate white identity and, simultaneously, the very Otherness it seeks to exclude" (2001: 1). In this construction, carnival is about spectacle and display and would include, *inter alia*, minstrel shows, the World's Fairs, New Coney Island fun fairs and the rest.

Rationalizing these and other uses of the term, Sam Kinser (1990: 4) suggests differentiating between usages applying upper case against those employing lower case. Spelling "carnival" with lower case refers to the generalized meaning of showy amusement. "Carnival" with a capital letter denotes something more specific and much closer to the focus of attention here. It stands for a non-ambulatory annual festival in the European tradition, but not one that is ossified. Its content generally involves visually spectacular displays and parades that proceed according to sets of rules and rituals that are embedded in place as well as in the social order. As such, this definition describes with equal facility the Mardi Gras festivals of the Gulf Coast cities, Mexican-influenced fiestas, St Patrick's Day parades, or gay and lesbian parades. At times, corporate sponsorship or themed marketing by the host city may produce commodified spectacle that may disguise Carnival's underlying cultural politics—at least, as far as the visitor is concerned. Despite this, Carnival retains a transgressive dimension that sets it apart from fairs or other occasions that produce parades, such as the Fourth of July, Thanksgiving or Christmas festivities held in towns and cities throughout the United States.

Making that point, of course, directly invites questions as to precisely what is meant by "transgression." For example, what in practice does transgression mean in the case of a Mexican fiesta? Are its practices inherently any more or less transgressive than a Gay Pride parade? And does the inevitable growth of ritualized practices numb the sense of standing outside of social norms?

To answer those questions, it is important to pay attention to three aspects of the cultural politics of Carnival—where "cultural politics" is taken to mean the field concerned with the power behind meaning and the way that the exercise of such power advances the position of particular groups and their interests (Barker 2000: 383). The first is self-conception. Frequently the notion of transgression and of the struggle associated with it is important to the way that the organizing group itself defines and represents the event. As such, the purpose of Carnival partly becomes an exercise in externalizing social memory, whereby the festivities and their organization express important ideas about the group's past. This conclusion applies even to recently established events that historians might lightly dismiss as invented tradition (Hobsbawm and Ranger 1983), provided that the event excavates sufficiently significant themes in social memory for participants to believe it expresses group identity. The second involves immanent possibilities. Even if an event seems long-established and essentially expressive of the existing social order, there remains within Carnival the possibility of transgression, and this can be readily mobilized for significant purposes as and when necessary. This can arise if the group feels that its identity and interests are under threat. This is particularly important when combined with the third consideration, namely, the continuing assertion of rights. Carnival everywhere involves joy, revelry, and an invitation to join in the festive mood. However, for a number of groups who have struggled to assert their rights in American society, Carnival may well gain or regain a harder edge—sometimes to the point at which the mood of celebration is replaced by something that is quite different.

US-American Carnival

Illustrations of these points readily appear when surveying American Carnival practices; some long-established and others of rather more recent origin. The St Patrick's Day parades, which take place annually on March 17, are a prominent example of the former (Abramson 1973, Marston 1989, Cronin and Adair 2002). Many have been an essential part of the cultural infrastructure of cities that have recognizable clusters of Irish Americans for several centuries, with the oldest being those held in Boston, which was first held in 1761 (although some claim 1737), and New York (1762). Eighteen cities had parades by 1890 and, moving towards the present day, one estimate suggests that there were at least 175 St Patrick's Day parades held in the United States during the 1980s (Kelton 1985: 93). Although these were predominantly Protestant parades at the beginning of the nineteenth century, their character changed dramatically over time, especially as the major wave of emigration took place after the Potato Famine of the 1840s (Moss 1995). Contemporary accounts speak of the parades as a vehicle for the new Irish and predominantly Catholic immigrant communities to revel in the freedoms to celebrate their culture openly and even to wear green—something that at times was proscribed by the repressive British regime in the old country (Cronin and Adair 2002).

The parades remain overwhelmingly festive occasions, with the marching bands, step dancers, people dressed as leprechauns, shamrockery, and the rest acting as the accompaniments for notably warm and hospitable celebrations. Nevertheless, their content has been periodically shaped and reshaped by the course of radical Irish nationalism. The Irish community in Seattle, for example, only founded their parade in 1972 as part of a response to what had happened in Northern Ireland two months earlier, when a civil rights march in Derry had left 13 protesters dead on what became known as "Bloody Sunday." In the early 1980s, the radicalization affected many of the parades when Irish American groups felt that their voice on Northern Ireland was being ignored, largely due to the Reagan administration's close relationship with Margaret Thatcher's government in Britain.

In 1982, for example, marchers processed down New York's Fifth Avenue silently carrying crosses and wearing H-Block armbands to commemorate the deaths of Irish republican hunger strikers in the Maze Prison—their silence, understandably, being in sharp contradistinction to the celebratory norms of Carnival. For their part the Ancient Order of Hibernians, the organizer of the New York St Patrick's Day Parade since 1838, has occasionally appointed controversial Irish republican figures to be its Grand Marshal. These included Michael Flannery in 1983 and Pete King in 1985 (Kelton 1985: 99–102). Flannery had just four months previously been acquitted of charges of gun-running for the IRA (Irish Republican Army). King was an active campaigner for civil rights in Northern Ireland and a fervent supporter of Noraid, long regarded by the British Government as a principal source of funds for militant republicanism.

   The Irish, of course, are not the only group using Carnival forms to make their voice heard and at times to represent their struggles in the face of the dominant discourse in US-American society: Black Americans are another. Kathlyn Gay's invaluable guide to African American celebrations lists no less than 109 separate festivals, almost half of which have celebrations that are recognizably Carnivalesque (Gay 2007). Some have a considerable pedigree. The Juneteenth festival, for example, which commemorates the day on which the last slaves in the South were emancipated (June 19, 1865), has been observed in Texas since 1866 (Taylor 2002). Most other Black American festivals, however, were the products of the awakening and flowering of black consciousness that began in the 1950s and 1960s. These include Harlem Week (New York City), founded in 1974; the Black Pride Festival (Washington DC, 1975); Chattanooga's Bessie Smith Strut (1981); and Milwaukee's African World Festival (1982). Other festivals celebrated by communities nationwide specifically relate to civil rights struggles: most notably, Martin Luther King Jr's Birthday (established in 1983) and Rosa Parks Day (2005). These are not protest rallies per se and indeed have a commemorative tone that reflects community pride, but contain constant reminders that such festivities help carry forward the spirit of the Civil Rights movement and the right to be seen and heard.

   Perhaps the most open displays linking transgression and revelry in the contemporary United States come from the gay and lesbian movement, where Lory Britt and David Heise (2001), for example, have charted the role of Lesbian and Gay Pride Days, *inter alia*, as a catalyst in turning "shame and loneliness into pride and solidarity." There are two potential dates given as to when this movement started to deploy Carnival forms. One was 1958 when the New Orleans Mardi Gras gained its first Gay krewe (carnival club), the Krewe of Yuga or "KY krewe" [sic]. The other was 1969 with the founding of the Gay Pride movement whose first "parade," the march on the Stonewall Inn, took place in protest against unconstitutional raids on gay bars in New York by the Police Department (Armstrong 2006).

   Gay Carnival practices spread rapidly from these beginnings. A recent schedule revealed 73 US-American cities listed as having Gay Pride festivals and parades for 2008 (GayCityUSA 2008). Emergent findings from research (e.g., Kates and Belk 2001) reveal the festival as a collective site of ritualized community resistance that raises awareness of social injustice and discursively informs social meanings in everyday life outside the festival. There is, of course, more at stake here than just issues of cultural politics. For the municipality, the parades are represented as Carnivalesque celebrations that are at once statements of civic identity and commercialized experience. For the gay community, they can also be vehicles used to highlight elements of active discrimination. New York's Gay Pride Parade, for example, consciously employs the same routing as that for the St Patrick's Day Parade. Given that the Ancient Order of Hibernians specifically bans overtly gay participants from its parade, one can see that the choice of that route is not wholly accidental (Marston 2002). More subtly, it was noticeable that the parade that celebrated the twenty-fifth anniversary of the March on Stonewall began not in Greenwich Village or even at the Stonewall Inn where it all began but at the United Nations building—thereby eliding protection of gay rights with more general issues of human rights (Manalansan 1995).

There is much more that could be said to exemplify the remarkable growth and diversification of Carnivals in the United States over the last half century, but it is important also to recognize their shared dimensions for the cities that host them. With the decline of manufacturing industries, municipalities perforce strive to diversify their economies, frequently turning to the cultural sector to generate new income. Carnivals are increasingly important in this respect. These events draw in participants from specific sections of the community, but are increasingly incorporated into the urban agenda as a commodified form of social heritage that can be represented as national or international visitor attractions. The sums at stake are significant, although it is impossible to find statistics that treat these events on any comparable basis. To take some examples, we may note that the 2008 San Antonio Fiesta had an annual economic impact of $284 million; the various St Patrick's Day parades currently generate around $4 billion of spending a year, with enormous contributions to the tourist revenues of cities like New York and Boston; and even a small-scale festival like the 2006 Gay Pride Festival in Hillcrest, California, yielded "just over $1 million," including ticket sales, vendor rents, parade entry fees, sponsorship fees, and beverage sales (Lewis 2007).

These are substantial amounts and, not surprisingly, the business of selling Carnivals becomes a significant task for tourist agencies, convention bureaus, and more generally for place promoters and marketers in city hall. Details of the relevant selling strategies lie beyond the scope of this chapter (e.g., Gold and Ward 1994, Hoyle 2002, Jeong and Almeida 2004). Common ploys, however, include presenting them as part of local tradition—invented or otherwise—and often with didactic overtones; selling through theming, whereby Carnival events are "sold" alongside other kindred attractions as a package (Derrett 2004, Lukas 2007); and offering what might be considered "pseudo-participation." This last strategy invites the spectator to watch not as passive observer but, in the classic manner of Carnival, as a participant by virtue of being there. Some observers might argue that this testifies to a process of decline, seeing events that once possessed vitality and meaning for a community being reduced to what Guy Debord (1983: 28) termed "pseudo-goods to be coveted." The municipal response in particular, they feel, may serve to rob the event of its deeper meanings. In this vein, for example, Philip McGowan argues that the American tendency is to confine Carnival: "to remove its subversive and transgressive nature and simultaneously to display it within purported educative or judicial frameworks for instruction of the homogeneous masses" (2001: 1).

This, however, is a thin reading of the Carnivalesque. Despite the inevitable convergences with the sanitized entertainment-and-pedagogy ethos of the Disney theme park and its commercial importance to city exchequers, the true Carnival retains a transgressive edge. Its practices may become ritualized—at least to the gaze of the outsider—but they remain both important parts of the cultural matrix of the host city and deeply expressive of urban cultural politics. Those elements were certainly ingredients in the processes of negotiation that underpinned the staging of the 2006 Mardi Gras in New Orleans in the face of extreme adversity.

Mardi Gras

On August 29, 2005, television audiences round the world watched with morbid fascination as a powerful hurricane approached the Louisiana coast. At first the reports were optimistic. Hurricane Katrina, measured as a Category 5 storm over the Gulf, had apparently lost strength and only measured Category 3 when it reached land in Southern Louisiana and Mississippi. These reports, however, proved an inaccurate guide to the damage to be caused by the hurricane's 225-kilometers-an-hour winds, torrential rains, and storm water surges. The flood protection system in New Orleans failed in 53 different places (Levitt and Whitaker 2009: 2). The storm flattened a land mass the size of Great Britain (Hirsch 2009). Eighty per cent of the metropolitan area was flooded, with some parts 4.5 meters under water. New Orleans itself suffered damage to property and infrastructure in excess of $200 billion, with over 1,200 of the estimated total of around 1,860 deaths occurring in the city boundaries (Congleton 2007: 5–6, see also Dyson 2006).

The disaster forensically exposed the racialized character of the residential geography of the amphibious city of New Orleans (Colten 2000, Kelman 2003). In common with other cities in the Deep South, the era of slavery required blacks to live in close proximity to their white owners. The resulting mixed residential pattern saw the social distance imposed by a rigid and racialized housing market serve in place of physical distance. In the twentieth century, legislation to achieve civil rights and racial equality has led to greater residential segregation based on economic factors and informal segregation (Spain 1979, C. Taylor 2009). Largely as a result, the poorest and mostly black sections of the community found themselves relegated to cheaper low-lying land that was previously shunned, due to the well-known but fatally ignored threats of river flooding and damage from tropical storms sweeping in from the Gulf of Mexico (Lewis 2003). Not surprisingly, these groups suffered a disproportionate loss of lives compared with other groups and had a similar over-representation amongst the New Orleanians evacuated to towns and cities throughout the South just before or during the disaster, with little immediate prospect of return (Hartman and Squires 2006).

Two questions arose concerning the dangers posed to the city's unique and vibrant mix, often identified as a quintessential example of the American melting pot (Holiday Spot 2009), that underpins the city's cultural life and is often regarded as its most notable attribute. The first was whether or not it was possible to rebuild the city's traditional cultural infrastructure and ethos. As Kevin Fox Gotham (2007: 2–4) noted, there were at least three different sets of opinion on the matter. One first saw the flood almost in Biblical terms, but without the consistency of moral judgment. Here, the old culture was deemed to have been destroyed, with the likelihood that the visitor could only encounter the "banalization" of the city, "a culturally empty place divested of authenticity and communal value" (2007: 2, see also McKernan and Mulcahy 2008). Another interpretation held that a reconstructed New Orleans would emerge with its festivals intact, but reduced to bland Las Vegas-style entertainment with the city "an artificial and contrived version of its old urban self" (Gotham 2007: 2). This vision would not in itself be entirely new, since racially motivated tourist promotional policies during the interwar period (Stanonis 2006) and after had already contributed the makings of that trajectory. The third and most optimistic interpretation viewed post-Katrina rebuilding as "helping to foster a new appreciation and rebirth of local culture" (Gotham 2007: 3)—once again a policy objective with antecedents (e.g., McKinney 2006).

The other and more urgent question was whether it was right to stage the Mardi Gras in 2006 so soon after an event that had wrought such devastation. Although 2006 would have been the sesquicentenary of its modern foundation, it was not as if there was a long unbroken sequence of carnivals to be protected. There had been 13 occasions since 1857 on which the annual Mardi Gras festival had not been held, such as the years during the two World Wars and, most recently, during the Korean War. Not surprisingly, therefore, there was a fierce local debate among the major local stakeholders about whether or not a Carnival should have taken place in such circumstances.

The battle lines were joined by complex coalitions that transcended simple binary oppositions of pro-Mardi Gras economic interests and anti-Mardi Gras community groups. On one side was arraigned a loose coalition that believed it was morally unacceptable to stage a festive Carnival just six months after an event that brought so much death, destruction, and dispossession. An organized anti-Mardi Gras movement developed, which drew support from community activists and from New Orleanians living in cities such as Atlanta and Houston to which they had been evacuated. Their case centered on the plight of the black working class and underclass, who had suffered disproportionately from the deaths and property losses caused by the disaster (Gold 2005, Nichols 2005). They argued in particular that the requisite time and resources should be devoted to rectifying the city officials' lamentable failure to craft a comprehensive plan to rebuild the city not just to ensure that the levees would stand against future storms but also to allow the far-flung New Orleans Diaspora to return (D.K. Taylor 2009).

Ranged against them was an equally loose coalition that said the show must go on. This group included influential residents, political leaders, and commercial interests extending from the New Orleans Marketing and Convention Visitors Bureau down to the souvenir stores. Their case was partly economic and partly place promotional. Mardi Gras, they noted, was then estimated to raise between $750–800 million a year in spending and yield $55 million in state, city, and parish revenues (Longman 2005). Hospitality and tourism, the city's largest industry with 85,000 workers, might be considered ideally placed to stimulate revival through bringing in revenues and returning levels of employment back to pre-Katrina levels. Moreover, the staging of Mardi Gras might be taken as symbolizing the indomitable spirit of New Orleans, saying clearly to the American nation and wider world that the city was back on its feet. In the words of local historian Arthur Hardy: "Being at Mardi Gras 2006 will be like being in Times Square on the first New Year's Eve after September 11" (quoted in Nichols 2005).

The pro-Mardi Gras campaign, however, also drew support from within the black community and other poorer sections of society who had collectively suffered most from the hurricane and its aftermath. For many, the idea of cancelling the Mardi Gras festivities would only have compounded the sense of loss. This included the many New Orleanians who attend the parades not just as casual observers but as family groups, gathering together each year at specific places along the routes of the parades to meet and share picnics with friends and renew acquaintances. It also included the members of the krewes that hold the annual balls and organize participation in the float parades. Specifically existing to meet Mardi Gras obligations, this indigenous organizational form (Islam, Zyphur, and Boje 2008) operates on a year-round basis. It brings responsibilities for fundraising to support preparation of floats, costumes, and the trinkets thrown to spectators and for commitment of time to undertake preparatory activities, but also gives members a sense of belonging to a shared cause and often a feeling of exclusiveness, given that membership is by application. Inevitably, these groups had mixed feelings, but 31 out of 34 krewes quickly agreed to participate. For example, Charles Hamilton, the president of the Zulu krewe—a largely African American marching organization that had more members in devastated eastern New Orleans than any other krewe—recognized that there were elements of his group that believed the Carnival would be a distraction from attending to more pressing needs. Nevertheless, he supported staging the Mardi Gras because "[i]t's not just about fun. It's tradition, something that makes our city what it is" (quoted in Nichols 2005).

In the event, the decision was made to stage the 2006 Mardi Gras on the basis of six rather than the normal eleven days of festivities, with progressive return to the pre-Katrina pattern in subsequent years. Despite the vicissitudes of 2005, Mardi Gras has steadily regained its good humored but edgily transgressive character and remains the defining feature of community life. Indeed, it has been reinforced in that role by the scandalous delays in rebuilding the devastated areas of housing; a policy of delay that only benefits the property industry. Mardi Gras has assumed profound symbolic significance for those former residents who have yet to return. As one noted in relation to the 2009 Carnival: "[I have] been gone since Katrina, but I have been *coming home* each year for Mardi Gras" (Mardi Grass New Orleans 2009, emphasis added). For that person and others, Mardi Gras provides a way in which the dispossessed can register their continuing case. Simply by being there, they can again participate in a community of which they were once part and, quite possibly, gain a vehicle for ensuring that they are not forgotten.

Conclusion

That finding supports the theme that has featured repeatedly in this chapter. Carnival is always developing, always adding new agendas, and yet stubbornly retaining its enduring character. From the early eighteenth century onwards, numerous groups within US-American society have appropriated this well-established cultural form and adapted its underlying strategies for their own needs and purposes. It may be that sometimes the transgressive edge may be blunted by the process of selling as it is by the changes that occur in the material circumstances of the groups who participate, but the possibility of transgression remains there as a force that can be revived should the need arise. If nothing else, Carnival remains as a vital ingredient in the lives of groups of people who annually see a reminder of who they are, where they came from, and frequently the distance that they still have to travel.

Works Cited

Abramson, H. J. 1973. *Ethnic Diversity in Catholic America*. New York: John Wiley.

Anon. 1875. Italy: a Dull Carnival—The Rule of the Mendicants. *New York Times*, [Online, March 1]. Available at: http://query.nytimes.com/mem/archive-free/pdf?_r=1&res=9C0CEED6143CE63ABC4953DFB566838E669FDE [accessed: September 17, 2009].

Armstrong, E. 2006. Movements and Memory: the Making of the Stonewall Myth. *American Sociological Review*, 71(5), 724–51.

Bakhtin, M. 1984. *Rabelais and His World*, translated by Helene Iswolsky. Bloomington: Indiana University Press.

Barker, C. 2000. *Cultural Studies: Theory and Practice*. London: Sage.

Barnes, N. 2000. Body Talk: Notes on Women and Spectacle in contemporary Trinidad Carnival. *Small Axe*, 7, 93–105.

Britt, L. and Heise, D. 2000. From Shame to Pride in Identity Politics, in *Self, Identity, and Social Movements*, edited by S. Stryker, T. Owens, and R. White. Minneapolis: University of Minnesota Press, 252–69.

Colten, C. (ed.) 2000. *Transforming New Orleans and Its Environs: Centuries of Change*. Pittsburgh: University of Pittsburgh Press.

Congleton, R. 2006. The Story of Katrina: New Orleans and the Political Economy of Catastrophe. *Public Choice*, 127(1–2), 5–30.

Cronin, M. and Adair, D. 2002. *The Wearing of the Green: A History of Saint Patrick's Day*. London: Taylor and Francis.

Debord, G. 1983. *The Society of the Spectacle*, translated by Ken Knabb. London: Rebel Press.

Derrett, R. 2004. Festivals, Events and the Destination, in *Festival and Events Management: An International Arts and Culture Perspective*, edited by I. Yeoman et al. London: Butterworth–Heinemann, 32–52.

Dyson, M. 2006. *Come Hell or High Water: Hurricane Katrina and the Color of Disaster*. New York: Basic Civitas Books.

Gay, K. 2007. *African-American Holidays, Festivals, and Celebrations: The History, Customs, and Symbols Associated with Both Traditional and Contemporary Religious and Secular Events Observed by Americans of African Descent*. Detroit: Omnigraphics.

GayCityUSA. 2008. *International Gay Pride Parades and Events Calendar* [Online]. Available at: http://www.gaycityusa.com/pride.htm [accessed: September 6, 2009].

Gilmore, D. 1998. *Carnival and Culture: Sex, Symbol, and Status in Spain*. New Haven, CT: Yale University Press.

Gold, J. R. and Gold, M. 2012. *Festival Cities: Culture, Planning and Urban Life since 1918*. London: Routledge.

Gold, J. R. and Ward, S. (eds.) 1994. *Place Promotion: The Use of Publicity and Public Relations to Sell Towns and Regions*. Chichester: John Wiley.

Gold, S. 2005. New Orleans Mardi Gras Plans under Criticism. *Los Angeles Times*, December 7, A11.

Gotham, K. F. 2007. *Authentic New Orleans: Tourism, Culture, and Race in the Big Easy*. New York: New York University Press.

Grant, L. 2004 *Carnivalitis: The Conflicting Discourse of Carnival*. Jamaica, NY: Yacos.

Harris, M. 2003. *Carnival and Other Christian Festivals: Folk Theology and Folk Performance*. Austin: University of Texas Press.

Hartman, C. and Squires, G. (eds.) 2006. *There Is No Such Thing as a Natural Disaster: Race, Class, and Katrina*. London: Routledge.

Hennelly, M. 2002. Victorian Carnivalesque. *Victorian Literature and Culture*, 30(1), 365–81.

Hirsch, A. 2009. (Almost) a Closer Walk with Thee: historical reflections on New Orleans and Hurricane Katrina. *Journal of Urban History*, 35(5), 614–26.

Hobsbawm, E. and Ranger, T. (eds) 1983. *The Invention of Tradition*. Cambridge: Cambridge University Press.

The Holiday Spot. 2009. *Mardi Gras in the Melting Pot* [Online]. Available at: http://www.theholidayspot.com/mardigras/origin.htm#pot [accessed: September 11, 2009].

Hoyle, L. 2002. *Event Marketing: How to Successfully Promote Events, Festivals, Conventions, and Expositions*. New York: John Wiley.

Humphrey, C. 2001. *The Politics of Carnival: Festive Misrule in Medieval England*. Manchester: Manchester University Press.

Islam, G., Zyphur, M.J. and Boje, D. 2008. Carnival and Spectacle in Krewe de Vieux and the Mystic Krewe of Spermes: the Mingling of Organization and Celebration. *Organization Studies*, 29(12), 1565–89.

Jackson, P. 1988. Street Life: the Politics of Carnival. *Environment and Planning D: Society and Space*, 6(2), 213–27.

Jeong, S. and Almeida, C. 2004. Cultural Politics and Contested Place Identity. *Annals of Tourism Research*, 31(3), 640–56.

Kates, S. and Belk, R. 2001. The Meanings of Lesbian and Gay Pride Day. *Journal of Contemporary Ethnography*, 30(4), 392–429.

Kelman, A. 2003. *A River and Its City: The Nature of Landscape in New Orleans*. Berkeley: University of California Press.

Kelton, J. 1985. New York City St. Patrick's Day parade: invention of contention and consensus. *The Drama Review*, 29(3), 93–105.

Kertzer, D. 1988. *Ritual, Politics, and Power*. New Haven, CT: Yale University Press.

Kinser, S. 1990. *Carnival, American Style: Mardi Gras at New Orleans and Mobile*. Chicago: University of Chicago Press.

Ladurie, E. 1980. *Carnival: A People's Uprising at Romans, 1579–1580*, translated by Mary Feeney. London: Scolar Press.

Levitt, J. and Whitaker, M. 2009. Truth Crushed to Earth will Rise Again: Katrina and its Aftermath, in *Hurricane Katrina: America's Unnatural Disaster*, edited by J. Levitt and M. Whitaker. Lincoln: University of Nebraska Press, 1–21.

Lewis, C. 2007. Hillcrest's Gay Pride festival, parade big boost for the neighborhood. *San Diego Business Journal*, 28(10), 1.

Lewis, P. 2003. *New Orleans: The Making of an Urban Landscape*. Santa Fe: Centre for American Places.

Longman, J. 2005. Mardi Gras to the Rescue? Doubts Grow. *New York Times*, November 26, 9,

Lukas, S. A. (ed.) 2007. *The Themed Space: Locating Culture, Nation, and Self.* Lanham, MD: Rowman and Littlefield.

McGowan, P. 2001. *American Carnival: Seeing and Reading American Culture*. New York: Praeger.

McKernan, J. and Mulcahy, K. V. 2008. Hurricane Katrina: a Cultural Chernobyl. *Journal of Arts Management, Law and Society*, 38(3), 217–30.

McKinney, L. 2006. *New Orleans: A Cultural History*. Oxford: Oxford University Press.

Manalansan, M. F. 1995. In the Shadows of Stonewall: Examining Gay transnational politics and the diasporic dilemma. *GLQ: A Journal of Lesbian and Gay Studies*, 2(4), 425–38.

Mardi Gras New Orleans. 2009. *Guestbook: August 25, 2009* [Online]. Available at: http://www.mardigrasneworleans.com/guestbook/ [accessed: September 17, 2009].

Marston, S. A. 1989. Public Rituals and Community Power: St. Patrick's Day parades in Lowell, Massachusetts, 1841–1874. *Political Geography Quarterly*, 8(3), 255–70.

Marston, S. A. 2002. Making Difference: Conflict over Irish Identity in the New York City St. Patrick's Day parade. *Political Geography*, 21(3), 373–92.

Mintz, J. R. 1997. *Carnival Song and Society: Gossip, Sexuality and Creativity in Andalusia*. Oxford: Berg.

Moss, K. 1995. St. Patrick's Day Celebrations and the Formation of Irish-American Identity, 1845–1875. *Journal of Social History*, 29(1), 125–48.

Nichols, B. 2005. Mardi Gras '06: "We've got to have this party;" New Orleans determined to go on with 150th-anniversary carnival. *USA Today*, November 21, A12.

Ravenscroft, N. and Gilchrist, P. 2009. Spaces of Transgression: Governance, Discipline and Reworking the Carnivalesque? *Leisure Studies*, 28(1), 35–49.

Scribner, B. 1978. Reformation, Carnival and the World turned upside-down. *Social History*, 3(3), 303–29.

Spain, D. 1979. Race Relations and Residential Segregation in New Orleans: Two Centuries of Paradox. *Annals of the American Academy of Political and Social Science,* 441(1), 82–96.

Stallybrass, P. and White, A. 1986. *The Politics and Poetics of Transgression*. London: Methuen.

Stanonis, A. J. 2006. *Creating the Big Easy: New Orleans and the Emergence of Modern Tourism, 1918–1945*. Athens: University of Georgia Press.

Tanner, T. 1992. *Venice Desired*. Cambridge, MA: Harvard University Press.

Taylor, C. A. 2002. *Juneteenth: A Celebration of Freedom*. Greensboro, NC: Open Hand Publishing.

Taylor, C. 2009. Hurricane Katrina and the Myth of the Post-civil Rights Era. *Journal of Urban History*, 35(5), 640–55.

Taylor, D. K. 2009. "Chocolate City:" Personal Reflections from New Orleans, August 29, 2006. *Journal of Urban History*, 35(5), 614–26.

Twitchell, J. B. 1992. *Carnival Culture: The Trashing of Taste in America*. New York: Columbia University Press.

Weeks, L. and Whoriskey, P. 2006. Masks Hide an Empty Spirit: New Orleans Revelry Goes On, But Something is Missing. *Washington Post*, February 28, A5.

Yates, W. 1997. An Introduction to the Grotesque: Theoretical and Theological Contributions, in *The Grotesque in Art and Literature: Theological Reflections*, edited by James Luther Adams and Wilson Yates. Grand Rapids: William B. Eerdmans, 1–68.

Chapter 2

"What Did I Do to Be so Global and Blue?"—Blues as Commodity: Tourism, Politics of Authenticity, and Blues Clubs in Chicago Today

Wilfried Raussert and Christina Seeliger

American Studies and Studies of Ethnicity

The internationalization of American studies may trace its beginnings back to the increased interest in African American Studies in the 1960s on both sides of the Atlantic; first in the United States, but subsequently also beyond its boundaries. The decades after World War II brought a steadily growing awareness of cultural and ethnic differences within the public discourse of the United States. The Civil Rights movement became a propelling force for establishing questions of racial and ethnic injustice as central political issues. Artistically as well as politically, minority groups voiced their claims in the turbulent decade of the 1960s. With the passage of the 1965 Immigration and Nationality Act major shifts occurred in the ethnic landscape of the United States. We witness an increasing diversity in terms of communities along the lines of religion, ethnicity, and class. The new complexities and dynamics emerging in the context of modern transportation, new means of electronic communication, transnational labor, and new (at times "anti-American") nationalisms have transformed not only the notion of cultural origin and new home but also the idea of America as multicultural society (Banerjee, Birkle, and Raussert 2006: 311).

Ethnicity in the Context of Global Commodity Culture

Multiculturalism, in the aftermath of the turbulent 1960s, has become a highly contested concept in various spheres such as politics and academia. More recent debates circling around multiculturalism have shown that "ethnicity" not only figures as a culturally and politically but also as an economically powerful signifier within processes of globalization. "Selling ethnicity," with a nod to the thematic focus of this compilation, is both frequent economic manifestation as well as powerful marketing strategy, and both phenomena have emerged from the omnipresence of multiculturalism in discourses within contemporary urban society. Various discourses intersect when the increasing interest in postcolonial studies, the academic institutionalization of ethnic studies, the ever-growing expansion of the tourist industry, as well as consumers' interest in ethnic products in terms of food, fashion and culture meet, collide, and fuse in global economic and cultural exchange. The processes behind "selling ethnicity" are manifold as are the questions about agency revolving around the processes of commodifying self and other. As black feminist scholar bell hooks argues, "[t]he commodification of otherness has been so successful because it is offered as a new delight, more intense, more satisfying than normal ways of doing and feeling. Within commodity culture, ethnicity becomes spice, seasoning that can live up the dull dish that is mainstream white culture" (1992: 21).

Stephen Foster reminds us that fully domesticating the exotic would limit its potential to create surprising effects, thereby assimilating it "into the humdrum of everyday routines" (1982: 21–2). Thus, while the exotic may be used to assimilate and neutralize cultural difference, it also signifies a comprehension of difference and diversity that is inevitably distorted and always uncontrollable. Hence, assimilation is necessarily limited and, as Graham Huggan points out, "as a system, then, exoticism functions along predictable lines but with unpredictable content; and its political dimensions are similarly unstable" (2001: 14). Moreover it is necessary to take into consideration that political as well as economic changes bring about shifts in agency that also influence who sells "ethnicity" where, when, how, and for what purpose beyond simply gaining profit. In this context, Huggan speaks of a whole "cosmopolitan alterity industry" that relies heavily on selling not only ethnicity, but "exotic myths" (2001: xiii). While the products of this industry are mainly geared for easy consumption, Huggan points out that the commodification of cultural difference is seldom only one-sided as it also, at least partly, aims to fulfill an educational function. But even though the alterity industry offers a space of agency to Huggan's "postcolonial exotic," its target groups are nevertheless "mostly, if not exclusively, ... the capitalist societies of the West" (2001: 68).

New Challenge for American Studies: Exploring the Intersection of Cultural and Economic Politics

"Selling ethnicity" then is one among numerous factors that highlight the mobility and fluidity of ethnicity today. Processes of globalization and transnational networking help produce and distribute notions and marketable adaptations of fluid identities and mobile ethnicities in transactions of global trade, in their impact on how culture is constructed, perceived, and distributed. However, they also pose a significant challenge for Cultural Studies and American Studies, the latter being closest to the scholarly endeavor to explore US-American culture as one of the most powerful global players in that they highlight the necessity to explore the intersection between economic and cultural politics for the development of new identities more fully. Further, they pose a challenge to critical views of ethnicity as a site of resistance as they have figured prominently in postcolonial and minority discourses. As Winfried Fluck reminds us in his reflection on resistance in new turns in the studies of the Americas:

> For some time, American Studies put all hopes for resistance on marginalized groups and ethnic subcultures, until the critique of essentialism destroyed the equation of disenfranchised minority groups with resistance and left only the idea of a negating potential of flexible, multiple identities. All of this is the result of an increasingly radical and sweeping power analysis. If systematic power is all-pervasive, the hope for resistance can only be placed in the margins of that system, and even if the margins can no longer possess a quasi inbuilt oppositional, then only a flexible identity can function as a resort of last hope. This new utopia is often space- or territory-based, for example in the emphasis on border zones, diasporas, or intermediate spaces, because, as the argument goes, such spaces force their inhabitants to adopt several identities and thus seem ideally suited to create models of resistance. (2007: 70)

The fact that ethnicity has been discovered by capitalist interest does not necessarily mean that ethnicity loses its potential to function as a site of resistance. Rather, Fluck's comments highlight that ethnicity has become a contested terrain between mainstream and margin. Most obvious, as we argue, such a competition becomes visible in the context of tourism and the tourist's desire to consume authentic otherness. The growth of cultural tourism and the expansion of the heritage industry aim at satisfying a growing longing for the archaic and authentic in times of rapid global changes. Tourist sites, in this context, are likely to be contested spaces because they are located at the intersection of competing economic, social, cultural, and political influences. Frequently tourist attractions are developed in collaboration between national and international companies and institutions and often lie outside the control of local residents. The latter may try to gain access to the tourist industry in their neighborhood and they may also develop strategies of resistance. In both cases questions about agency emerge addressing issues such as who will gain access to the market and who will shape the development of the heritage industry (Selwyn 1996, Tucker 1997, Low and Lawrence-Zúñiga 2003).

The Contemporary Chicago Blues Scene: Politics of Authenticity and Chicago as "Home of the Blues" City

In this chapter we will focus on the blues as vital part of contemporary commodity culture, especially with regard to recent tourist marketing strategies presenting the city of Chicago as "Home of the Blues." The blues represents a particularly interesting case study since as musical expression it spans a long history of changes encompassing a leading role as secular vernacular expression in the Harlem Renaissance, a role as continuous backbone for innovations in jazz throughout the twentieth century, a role as major influence on the evolution of rhythm and blues, rock n' roll, and rock music in the sixties and beyond, and as universalist language to express pain and catharsis. In addition, the blues plays a role as background music in commercials and as recent tourist attraction for those who seek "their" black American authenticity in the context of music entertainment. To explore the intersection of cultural and economic politics we like to draw upon a series of questions guiding our inquiry into Chicago's contemporary blues scene: What are the politics inherent in the attribution of authenticity within the marketing strategies for blues music? How is black authenticity constructed, represented, and marketed with regard to blues music as black cultural heritage and tourist industry? What are the differences between the tourist conceptions and historical African American conceptions of blues music? How is agency distributed locally and how does it shift in the context of changing national politics and a growing global tourist industry?

Recent Developments in the Institutionalization of Blues Music as Cultural Heritage and Tourist Attraction

The marketing of Chicago as a city of ethnic diversity and authenticity is a propelling force behind the process of turning blues heritage and music into a vital part of Chicago commodity culture. Recent studies of urban spaces have emphasized postmodern developments of turning cities into sites of illusion and consumption. Christine Boyer (1994) refers to the "city of illusion," Charles Ruthesier (1996) labels the city "a non-place urban realm," and Sharon Zukin (1995) writes about the "city of cultural consumption." Such conceptions of the city are based upon Frederic Jameson's (1991) notion of postmodern urban spaces as a product of late capitalism. Frequently the marketing strategy of turning the city into a space of consumption is already visible at the gateways into the cities. Chicago's Hare International Airport represents one such site at which the process of establishing Chicago as home of the blues lures the arriving traveler and tourist into visiting the various sites of blues heritage and culture, as they have emerged in tourism and recent urban planning in Chicago. The walls of the hallway between arrival gates and immigration are decorated with enlarged photographs and paintings of musicians and performances, and tabloids with historical data about music create an immediate aura of blues as an inseparable core identity of the city of Chicago. As Marc Augé puts it, airports along with superstores and railway stations are non-places that "do not contain any organic society" (1995: 112). They are sites of coming and going and invite travelers and tourists into a realm of "translocality:" new conceptions of space, identity and commodity can flow freely within. The Chicago Museum shop at one of the crossroad sections of Hare International Airport makes it possible to consume the already installed aura of the blues city in forms of books, CDs, DVDs, photographs, and postcards and invites the traveler and tourist to purchase the blues spirit captured in recordings by home-based artists such as Koko Taylor, Carl Weathersby, and Junior Wells.

A series of records released by "Big Chicago" Records, including collections such as *A Chicago Jazz Tour, A Chicago Blues Tour, Hidden Treasures: Irish Music in Chicago*, and *The Chicago Music Scene*, underscore the culture-commodity, ethnicity-commodity link infusing the city's self-positioning as tourist attraction. As the example of *The Chicago Music Scene* illustrates, the music scene is ethnically defined and divided into Latin Mambo, Irish Traditional, African American Blues, and Jazz as global blackness. In addition to a selection of representative artists for the various genres the CD booklet also provides a city map, aptly named "club locator map," pointing towards city-sponsored music venues and helping tourists to navigate their way through Chicago night life.

Once tourists have found their way into the heart of downtown Chicago, programs sponsored by the Chicago Office of Tourism and the Department of Cultural Affairs, such as Chicago Neighborhood Tours (CNT), offer bus tours departing from the Chicago Cultural Center to numerous ethnic enclaves like the Jewish commercial areas on Devon Avenue, Puerto Rican neighborhoods in Humboldt Park, and blues heritage sites of the South Side's Bronzeville District. As David Grazian aptly puts it, "by capitalizing on exoticized notions of ethnicity, community, and urban space, these "motorcoach excursions" transform local communities and their landscapes of everyday social interaction into commodified tourist attractions" (2003: 201).

Not surprisingly then the blues scene presented to tourists frequently differs from the contemporary local music clubs. The tourist agenda of Chicago Neighborhood Tours appears eager on accommodating the tourist's quest for black authenticity. When the Chicago blues tour bus hits selected blues joints one often encounters only black performers on the band stand, and their repertoire focuses on blues classics from *Hootchie Cootchie Man* to *Crossroad Blues* during special tourist nights. What is addressed is the tourists' yearning for familiar sounds and well-established images embedded in a collective exotic blues imaginary. This imaginary is also fed with the unfortunate yet persistent power of racially charged and essentialist ideas of black sexuality. Skin color, run-down juke joints, and an ambiance of poverty frame the process of "blackening" the tourists' experience of racial difference. If the same tourists may accidentally return to the club another night, outside a regular motorcoach excursion, they would most likely encounter a different setting: a suburban all-white band performing recent local blues numbers emerging from current suburban experiences in Chicago.

Chicago's Creation as "Authentic Blues City" and the Transnational Tourist

The Chicago blues tours guide the tourists toward a mix of venues that taken together help fabricate and propagate the image of a blues city. Within such processes of fabricating we have to remember, as Richard A. Peterson claims, that the authentic is "a socially agreed upon construct in which the past is to a degree misremembered" (1997: 3). Not surprisingly reconstructions of cultural memory and new creations overlap in building Chicago as blues city. In 1984 the first Chicago Blues Festival was launched and has become an annual attraction for blues fans, local and global. Naturally the music festival is hailed as a milestone for preserving Chicago's African American heritage annually. While the clubs and bar venues focus on the performance of live music, institutes such as the Chicago Blues Archives as part of Chicago Public Library and the Chicago Blues Museum institutionalize the blues for the scholarly and public gaze. While the blues museum initially had its mobile home in a tent during the early Chicago Blues Festivals, it has found a permanent home in a 100-year-old former warehouse in an industrial park in Bridgeport, and it features showcases with photos, memorabilia, music instruments, and live performance film footage of blues and jazz artists from the 1920s to the present (cf. Hanson 2007: 70–71). The displays are arranged in a multimedia manner so that visitors are able to learn about the blues in its cultural and historical context. Next to the blues museum, there are stops at gravesites, music studios like Chess Studios, and the Jazz Record Mart. With regard to blues clubs advertised, one frequently encounters an overemphasis on fetishization to recreate a mythic, rural southern ambiance for the blues setting. The Voodoo Lounge at the Redfish, for instance, poses as black southern through "Mardi Gras masks, voodoo dolls, cartoon alligators, and other faux swamp kitsch" (Hanson 2007: 32). The Backporch Stage at the House of Blues presents itself as blues ambiance through Hoodoo candles, folk art painting, and bas-relief portraits of blues legends from Charley Patton to B.B. King (cf. Hanson 2007: 9). Often these places look like pop-culture replica of real life. These clubs then combine multimedia in their decoration to create a nostalgia-based image of blues ambiance.

While it is true that the blues experienced earlier days of commodification in the heydays of the Harlem Renaissance with the success of Bessie Smith and Ma Rainey, whose recordings with Race Records sold immensely and whose tours were propagated with great success not only in the south of the United States, the new phase of commodification is different. Whereas black consumers were mainly addressed and reached through Race Records in the 1920s, the blues bus tours of today and many of the blues clubs in Chicago aim at catching the attention of a transnational audience comprising Japanese, German, and French tourists. Certainly one of the reasons for this shift is that blues music no longer holds the stronghold position among secular black musical forms as it did in the 1920s. Moreover contemporary conceptions of what blues is differ quite succinctly between black and white audiences. "The blues in the African American community often incorporates soul, funk, jazz, and R&B," Karen Hanson (2007: 8) reminds us. The blues is not only seen as one among many different musical idioms within the African American cultural heritage; it is also seen as a developing and changing expression or part of what James A. Snead describes as a process of repetition with variation (1998: 73). What the "transnational tourists" (we are aware of the generalist reduction this expression may contain), as we call them, look for in blues is an expression of African American authenticity that is based on racial as well as aesthetic expectations. Hence the transnational tourist expects to find all-black blues bands, and, aesthetically, they long to listen to blues classics such as *Hootchie Cootchie Man* in a performance style that is generally described as raw, unpolished, and highly expressive. While the blues archive and the blues museum intend to preserve the memory of the origins and developments of blues in a historical perspective, the transnational tourists' exotic expectations tend to confine the blues within a realm of static black musical expression. In that way the transnational tourists are also different from the local listeners in Chicago's suburban blues bars, in which frequently all-white blues bands as well as ethnically mixed bands perform original song material for their local audiences, and in which open jam sessions with audience participation are regular features of the performances.

Politics of Authenticity: Politicization and Commodification of Blues Music

The ways blues clubs are decorated in a pop culture replica fashion and the fact that frequently all-black bands are hired to perform on tourist nights raise questions: What is the role of authenticity in the blues tourist industry and how is authenticity fabricated in the commodification of blues music today? Walter Hill's film *Crossroads*, released by Columbia Pictures in 1986, illustrates an important aspect of politics of authenticity in the context of black music and debates about cultural proprietorship. The encounter between Eugene Martin, a white student of classical guitar at the Juilliard School of Music, and the African American master of the blues harmonica, Willie Brown, in a prison hospital reenacts a frequent call for a re-authentification of black music. In a moody and sardonic way Willie Brown initially dismisses Eugene's enthusiasm for blues and his search for the one lost song by blues legend Robert Johnson. He mocks Eugene as a want-to-be bluesman from Long Island, ridiculing his privileged social condition as a white college boy. He also doubts Eugene's musical talent to play the blues—"you are lacking mileage, what you are playing sounds like 'bird-shit'"—and accuses him of art piracy—"you just want to be another one of those white musicians ripping off our musical heritage" (*Crossroads* 1986: 18:40–19:10'). In his critique of Eugene Martin, Willie Brown at the same time fabricates a blues authenticity based upon race, place, and class.

Authenticity itself, David Grazian reminds us, "is never an objective quality inherent in things, but simply a shared set of beliefs about the nature of things we value in the world" (2003: 12). As Grazian explains,

> authenticity ... is always manufactured: like life itself, it is a grand performance, and while some performances may be more convincing than others, its status as a contrivance hardly changes as a result. ... Like other kinds of stereotypes, images of authenticity are idealized representations of reality, and are therefore little more than collectively produced fictions. (2003: 11–12)

When Willie Brown signifies upon the conception of the ideal bluesman he draws upon collective memory, embracing myths about the origin of blues as well as the legend of Robert Johnson's encounter with the devil at the crossroads. While Willie Brown's narrative highlights that authenticity is fabricated, it also reveals that authenticity is always created for certain purposes and that its fabrication is driven by desire. For Willie Brown this serves to protect cultural heritage. In his reclaiming of the blues for black musicians authenticity is an expression of proprietorship in the first place, with racial, class-related, and geographic belonging determining membership as well as musical capability. Authenticity hence becomes a means of self-assertion and claiming one's place in the cultural spectrum of the United States. Thus Willie Brown's narrative also riffs on at times radical forms of black self-assertion during the black nationalist period of the 1960s, when critics and poets such as Amiri Baraka created the image of African Americans as "blues people" (cf. Baraka 1963) and excluded whites from both the process of participation in and reception of African American art forms. In his narrative about geographic origin Willie Brown fabricates blues authenticity. He claims the blues as African American property and locates its cradle in the Mississippi Delta, symbolically protecting it from white intrusion when he refuses to reveal his knowledge about the lost Robert Johnson song to Eugene. At the same time the story shows his awareness of the synergistic relationship between culture and commerce, because a recording of the one lost Robert Johnson song would turn it immediately into a part of commodity culture.

Numerous critics in recent years have lamented the increasing commercialization of blues music as tourist industry. John Miller Jones, scholar, poet, and musician born in Memphis, expresses such critique poetically in his "The New Beale Street Blues:

> Furry's dead and gone but got his golden sidewalk note,
> And old W.C. still stands tall in Handy Park.
> And B.B.'s "Lucille" can now and then be heard.
> But at the Pantaze Drug Store—where you never paid more, often less
> Beale Street ain't talkin' no more ...
> There's a café there that claims to be a piece of rock
> Iconic Elvis everywhere on a matchbook or a clock.

The fans come from Saipan, Japan and even Germany, too.
For downhome, kickass blues and some genuine barbecue.
But it all looks too cute, Beale Street seems to have gone mute.
Cause Beale Street's gone global and its blues now sounds too modal.
You'll meet Japanese and Taiwanese and Teutons and Celts
They're buyin' Elvis caps, T-Shirts, tank tops and belts … . (2008: 7)

The tourists travelling to Memphis's Beale Street in Jones's poem crave for the genuine blues experience. The fact that they end up buying Elvis caps and other paraphernalia shows how cultural products once valued for their authenticity turn into mere souvenirs and commodities. Ironically, the increasing process of commodifying traditional music, arts, and food seems to lead to an intensified search for authenticity at the same time. The narratives describing the symbols of authenticity in the context of tourism and contemporary local blues clubs differ only slightly from Willie Brown's in that they favor lower class all-black bands and ramshackle juke joints. With reference to the local blues club in the collective public imagination, David Grazian comments sardonically that clubs are considered genuine only if:

> [t]hey hire authentic looking blues musicians, who are generally uneducated American black men afflicted by blindness, or else they walk on a wooden leg or with a secondhand crutch; as they are defiantly poor, they drive beat-up Fords and old Chevy trucks, and usually cannot read or write their own name. Their audiences are usually black as well, with the occasional white customer surfacing only if they are sufficiently old, poor, drunk, or blind. (2003: 13)

Why in the first place do tourists from Japan, Germany, or Russia crave for an authentic blues setting? The fact that they do is obvious, as is the fact that the tourist market responds to such desire and caters music tours accordingly. A number of factors coincide and, as we argue, a fast-paced, increasingly globalized world that imposes changes on local as well as global communities perpetually intensifies the longing for so-called authenticity. Authenticity, in this context, signifies relic and a desire for origin as well as a longing for the real in what Baudrillard labels "simulacrum," a world that constantly fabricates copies and prefers the virtual to the real. That what Walter Benjamin has called the "aura" of the artwork seems long lost within a culture industry that favors reproduction, reproducibility, and repetition for the purpose of selling. But, more importantly, we argue, the strongly-felt loss of stability and continuity in times of transnational migration and globalization in places that were once called home in a static sense has created an ever-increasing urge to find the authentic some place else. Especially in context of performances, not only the content of the performance seems to play a role for the perceived authenticity, but also, crucially, the identity of the performer. As Timothy Brennan remarks, "exoticism" is more and more used "in both does and is" (1997: 115)—the touristic demand thereby overlaps interestingly with the culturally radical position of, for example, Willie Brown in the example above. However, the apparent demand for authenticity might not just be an indicator of touristic naiveté but could also be seen as a willing acceptance of the deceit or a willing suspension of disbelief. As Graham Huggan points out, it is "paradoxically ... the very *falseness* of nostalgia that makes it all the more appealing" (2001: 179), as the "inauthenticity of nostalgia is its own motivating force" (Huggan and Holland 1998: 20). While authenticity, to cultural critics such as Matthew Arnold, signals culture as juxtaposed to and superior to manufacture, thus becoming an essential quality standard for and within the overall perfection of society, authenticity, in the mind of the contemporary tourist, has turned into a quest for the static and the nostalgic. With regard to ethnicity, this search is often accompanied by idealized and exoticized visions of unchanged and still functioning communities.

New Agency in the Process of Commodification: Black Cultural Politics in Chicago

One of the propelling forces behind the creation of the Chicago Blues Festival in 1984 was the election of Harold Washington as Chicago's first African American mayor in April 1983. Just one day after his election, on April 30, famous blues composer and guitarist Muddy Waters passed away. The dynamics of celebration and mourning within Chicago's African American community helped spur a call for paying tribute to blues music as major manifestation of black cultural expression within the United States and as a signifier for Chicago's urban identity as cultural hub. A growing black political agency due to Harold Washington's election and a steadily increasing presence of black culture within the public spheres of the United States coalesced in this early process of institutionalizing the blues as both cultural heritage and tourist attraction. "The one-stage festival was scheduled for June 8, 9, and 10 and was dedicated to the memory of Muddy Waters. To open the festival Mayor Washington signed an official proclamation extolling the blues as 'the heart and soul of the Chicago experience'" (Hanson 2007: 55). By now, more than twenty-five years later, the "heart and soul" has turned, though little as compared to more popular musical forms of contemporary youth cultures, into a blues industry comprising blues clubs, blues archives, a blues museum, and blues tourist bus tours. Hence we can distinguish a process of institutionalization but with a clear shift in agency. African American cultural politics, to a large extent, account for institutionalizing and preserving African American cultural heritage. Special blues sessions offered for parents and children also illustrate that educational purposes are part of the process of institutionalization and show that blues has moved from cultural marginalization into the center of American Cultural education. While the blues tours play with the desire for the exotic and do not necessarily support the neighborhood policies necessary to rebuild structures for communal living, the process of commercialization has offered both black and white blues musicians opportunities to partake in processes of propagating, selling, but also extending the blues traditions into something new. While the tourist blues clubs may thrive on nostalgia primarily and present the blues as static and authentic expression of the past, it is in the suburbs of Chicago where local blues clubs and their musicians profit from Chicago's self-created image of home-of-the-blues city. And while blues may no longer be explicitly "black," it prospers as developing and changing musical form that gives new creative expression to contemporary urban experiences among white and/or multiethnic communities.

Reconsidering American Studies' Scholarly Perspectives on the Study of Ethnicity

As the example of blues in Chicago illustrates, a new focus on agency within the process of constructing, negotiating, and selling ethnicity is necessary if we want to comprehend the distribution and shifts of cultural politics within a transnationally embedded contemporary United States society. For contemporary American Studies scholarship this requires also to rethink concepts of multiculturalism, ethnic difference and how ethnicity figures within processes of political and economic redistribution. Graham Huggan warns that "academic concepts like postcolonialism are turned … into watchwords for the fashionable study of cultural otherness" (2001: vii). Academic postcolonialism thus has often "capitalised on its perceived marginality while helping turn marginality into a valuable intellectual commodity" (2001: viii) and has "taken full advantage of its own semantic vagueness" (2001: 1). Huggan therefore proceeds to distinguish between "postcolonialism" as "an anti-colonial intellectualism that reads and valorises the sign of social struggle in the faultiness of literary and cultural contexts" and "postcoloniality" as "a value-regulating mechanism within the global late-capitalist market of commodity exchange" (2001: 6). However, Huggan does not conclude that any scholarly study of ethnicity and multiculturality must be located firmly in the field of more academic postcolonialism. Instead, he emphasizes the interconnectivity of these processes by pointing out that academic research can no longer be untangled from industrial marketing processes, a fact that makes it the prerogative for scholarly research "to lay bare the workings of commodification" (2001: 64).

In a self-reflective and critical way, we as scholars, too, may have to give up a certain nostalgia for the authentic and stop chasing an abstract aesthetic ideal escaping socioeconomic reality, as this would limit the critical endeavor to understand the changes of musical forms such as the blues as part of an ongoing process of shifting ethnic markers, of changing political and economic agency, and of the migration of sounds that has taken on new speed in times of globalization and transnationalization. To conclude, let us return to the pun on words and the question in the title of our article: *What Did I Do to Be so Global and Blue?* The blues has gone white, Asian, and, one might say, global. But selling, as we argue, does not necessarily mean selling out. And while the process of commodification always also signifies an immersion in liberalist-capitalist marketing strategies, it allows for shifts in agency and thus also opens venues for self-representation and self-affirmation. While the new Chicago Blues may be white, many of the institutional forces to preserve and propagate the blues as African American art form in Chicago have shifted from white to black.

Works Cited

Augé, M. 1995. *Non-Places: Introduction to an Anthropology of Supermodernity.* London: Verso.

Banerjee, M. Birkle, C. and Raussert, W. 2006. Introduction. *Amerikastudien / American Studies*, 51(3), 311–21.

Baraka, A. 1963. *Blues People: Negro Music in White America.* New York: W. Morrow.

Boyer, C. 1994. *The City of Collective Memory.* Cambridge: MIT Press.

Crossroads (dir. Walter Hill, 1986).

Brennan, T. 1997. *At Home in the World: Cosmopolitanism Now.* Cambridge: Harvard University Press.

Fluck, W. 2007. Theories of American Culture and the Transnational Turn in American Studies. *REAL: Yearbook of Research in English and American Literature*, 23, 59–77.

Foster, S. 1982. Exoticism as a Symbolic System. *Dialectical Anthropology*, 7(1), 21–30.

Grazian, D. 2003. *Blue Chicago: The Search for Authenticity in Urban Blues Clubs.* Chicago: University of Chicago Press.

Hanson, K. 2007. *Today's Chicago Blues.* Chicago: Lake Claremont Press.

hooks, b. 1992. *Black Looks: Race and Representation.* Boston: South End Press.

Huggan, G. and Holland, P. 1998. *Tourists with Typewriters: Critical Reflections on Contemporary Travel Writing.* Ann Arbor: University of Michigan Press.

Huggan, G. 2001. *The Postcolonial Exotic: Marketing the Margins.* London: Routledge.

Jameson, F. 1991. *Postmodernism, or, the Cultural Logic of Late Capitalism.* New York: Verso.

Jones, J. M. 2008. The New Beale Street Blues, in *Traveling Sounds: Music Migration, and Identity in the US and Beyond*, edited by W. Raussert and J. M. Jones. Münster: Lit Verlag, 7.

Low, S. M. and Lawrence-Zúñiga, D. (eds.) 2003. *The Anthropology of Space and Place.* Oxford: Blackwell.

Melly, G. 1970. *Revolt into Style: The Pop Arts.* London: Penguin.

Peterson, R. A. 1997. *Creating Country Music: Fabricating Authenticity*. Chicago: University of Chicago Press.

Ruthesier, C. 1996. *Imagineering Atlanta: Making Place in the Non-Place Urban Realm*. New York: Verso.

Selwyn, T. 1996. *The Tourist Image: Myths and Myth Making in Tourism*. Chichester: John Wiley.

Snead, J. A. 1998. Repetition as a figure of black culture, in *The Jazz Cadence of American Culture*, edited by R. O'Meally. New York: Columbia University Press, 62–81.

Tucker, H. 1997. The Ideal Village: Interactions through Tourism, in *Tourists and Tourism: Identifying People and Places*, edited by S. Abrams, J. Waldren, and D. Macleod. Oxford: Berg, 91–106.

Weber, M. 1978 [1909]. *Economy and Society: An Outline of Interpretive Sociology*, vol.1. Berkley: University of California Press.

Zukin, S. 1995. *The Cultures of Cities*. Oxford: Blackwell.

Chapter 3

Insurrection and Symbolic Work: Graffiti in Oaxaca (Mexico) 2006/2007 as Subversion and Artistic Politics

Jens Kastner

For the Argentinean-Mexican Cultural Studies scholar Néstor García Canclini, graffiti are, next to comics, a constitutive expression of "hybrid cultures." As a constitutionally impure genre, that is, a genre oscillating between art and everyday praxis, graffiti indicates a fundamental transformation in contemporary cultures. García Canclini maintains that from now on artistic practices will purportedly dispense with "consistent paradigms" (2005: 243), high culture and popular culture veritably slide into one another, symbols of elite culture and mass culture blend together, and contemporary cultures are distinguished by principally "hybrid cultures"—according to García Canclini's award-winning diagnosis of the times. The German Cultural Studies scholar Andreas Hepp argues that for García Canclini, graffiti also exemplify "forms of communication, along which the symbolic production of meaning takes place" (2009: 168) by linking together visual and literary forms. What graffiti and comics also have in common, in addition to their genre-crossing references, is their omnipresence and mass distribution in urban spaces. Whereas comics can be received and consumed by a mass audience more effectively than any artistic work due to their presence in newspapers, in García Canclini's interpretation graffiti is also a symbolic way of taking possession of public space by inscribing it. With his thesis of the appropriation of public space and thus implicitly also the enabling of a political space through murals, García Canclini anticipates a set phrase that is frequently invoked, not only in subculture research.[1] Does it apply to every mural?

In the city of Oaxaca in Southern Mexico, capital of the federal state of Oaxaca, an uprising determined political, cultural and social life in the second

---

1   "All those who who apply graffiti claim a piece of space for themselves, in which they can express themselves" (Baeumer 2009: 112, translation mine). Simply through this occupying of space, claims to power and property are said to be called into question, regardless of the contents of the slogans/images and the intentions of their creators, Tobias Baeumer currently asserts in reference to the German context.

half of the year 2006.[2] After the uprising, which enjoyed broad support among the population, was brutally crushed in late November with the assistance of the federal police, one of the first measures taken by the authorities was to paint over all graffiti and traces of street art on the walls of the historic city center and beyond (cf. Collective Reinventions 2009: 145). Was the political wall painting too subversive?

Obviously this was a replay of the conflict over the (re)appropriation of urban space, as it had been discussed over the past few years in research and debates on urban sociology and sociology of space (cf., for example, Löw 2001, 2008). However, the fact that not every expression on the walls of public space first constitutes it as emancipatory or intervenes in it subversively is particularly evident in Mexico, where political wall art (muralism) has also served less subversive functions, such as supporting the state and attracting tourism. Nevertheless, these painted walls also bear witness to the history of the involvement of artistic practices in those social movements—a history that has been largely ignored by academic art history up to the present. Contrary to this ignorance and equally counter to a leftist euphoria that deciphers acts of creative subversion in every mural, the intention here is to set out in search of criteria for what is subversive in art. Repression, such as the authorities painting over pictures, can hardly serve as a criterion for distinguishing subversion. Suitable criteria are more likely to be found, such is the thesis of this article, in answering the question of how forms of "insurrection" (Negri) are linked with the battle over the "collective unconscious" (Bourdieu).

---

2   Whereas the events were hardly present in the international press, they were followed with interest by leftist groups, blogs and initiatives all over the world and celebrated in part from the beginning. The subtitle of the chronicle by Diego Enrique Osorno (2007) sounds paradigmatically promising: "The First Uprising of the 21st Century." The movement thus met with international sympathy "among all who resist the status quo," as the Collective Reinventions (2009: 135) noted in its extensive and critical appreciation. Not least of all due to the very specific local conditions of its emergence—and possibly also due to the limitations of the joint demands, among which the demand for the resignation of the governor was central—the insurrection movement in Mexico hardly had an impact beyond the state boundaries. And this happened despite national mobilizations, which followed as protests against the election of the conservative president Felipe Calderón from the Party of National Action (PAN) in the same year. Although relatively small actions of solidarity with the insurrectionists were also held in Berlin, Vienna, and other German-speaking cities, up to the present Oaxaca still remains an undiscovered and hence unexplored terrain for research in Germany on social movements. In Mexico, on the other hand, several books have meanwhile been published on the subject (an overview of publications so far is provided by Victor Raúl Martínez Vásquez in the introduction to the anthology he edited [cf. Martínez Vásquez 2009]).

The "Commune of Oaxaca" and the Artes Plásticas

The events that have been discussed as the "Commune of Oaxaca"[3] began with a strike of the teachers' union in May 2006. When the protest tent city in the Zócalo, the main square of the southern Mexican city—so popular among tourists— was brutally cleared by the police on June 14, large sections of the population expressed solidarity with the concerns of the teachers. On June 17 the *Asamblea Popular de los Pueblos de Oaxaca* (Popular Assembly of the Peoples of Oaxaca, APPO) was founded, in which roughly 350 organizations, groups, and initiatives joined together. Until officers of the federal police crushed off the Assembly in late November 2006, the movement against the authoritarian ruling governor Ulises Ruiz Ortiz from the *Partido Revolucionario Institucional* (Institutional Revolutionary Party, PRI) fought for and suffered through the most diverse highs and lows: a number of people, at least twenty-three, were shot by paramilitary groups close to the government—the most prominent victim was the US-American Indymedia journalist Brad Will in late October—, countless people were wounded, and about three hundred arrested. Sometimes as many as 1,500 barricades were counted in the picturesque historical city center, several radio stations were occupied, and a group of women even occasionally operated a television station that had previously been occupied.[4]

Along with the protagonist role of women and the tremendous involvement of organizations of the indigenous and the striking teachers, artists were especially noticeable as a particular group, when it came to answering the question of who were the subjects of this social mobilization. The participation of visual artists in the Oaxaca insurrection of 2006 was particularly conspicuous, as many of them placed their artistic practices directly in the urban space (cf. Lache Bolaños 2009, Nevaer 2009, Porras Ferreyra 2009).[5] Several artists working in the area of visual

---

3   During the uprising, the leftist Mexican daily newspaper *La Jornada* already saw the spirit of Louise Michel, the activist of the Paris Commune of 1871 who presided over the nighttime barricades (cf. Beas Torres 2006). Even though words, images, and deeds of the movement have generally been appreciated as a mobilization event, the question of whether it is possible to speak of a "Commune of Oaxaca" as an operative counter-model to the state-capitalist organization of society is highly controversial. In concurrence with the previously cited appraisal of the Collective Reinventions published in German in the journals *Die Aktion* (Hamburg, vol. 214, 2008) and *Kosmoprolet* (Berlin, vol. 2, 2009), it is probably to be understood as more of a goal "that the movement was striving for and, in the worst case, as mere wishful thinking" (Collective Reinventions 2009: 143).

4   A detailed chronology is found in the appendix of the book *Oaxaca sitiada* ("Occupied Oaxaca") written by the journalist Diego Enrique Osorno (2007) about the insurrection.

5   However, it is by no means the case that all artists took the side of the insurrectionists. Jaime Porras Ferreyra (2009: 231) also names a fraction of artists who were directly or indirectly connected with the regime. Among the artists in solidarity were not only representatives of the "artes plásticas," but actors from the fields of music, video, and

arts (*artes plásticas*) like Ana Santos or the group Arte Jaguar, for example, had already entered the public sphere before 2006 with their *arte urbano* (urban art). Turning to artistic productions such as graffiti and street art, for instance, it should first be emphasized that the artistic works created directly in conjunction with the movement are to be treated as specific cultural practices—that is, as art—not only for chronological reasons, but also for most of the other reasons that continue to make artistic works distinguishable from other kinds of objects and practices. On the one hand these include, in addition to the material preconditions, the appraisal from art criticism and the museum, in other words the capability of being exhibited,[6] preconditions for which are also their formal and possibly content-related references to works consecrated by art history. On the other hand, the works discussed in the following merge neither with the organizational forms nor with any other propagandist means of the movement. Nevertheless, because they not only have an obvious relationship to political forms of expression but are ultimately also—as will be argued in the following—themselves political forms of expression, Jaime Porras Ferreyra (2009) is right in describing the artistic practices as a long ignored challenge for Political Science. Yet what is it that now leads to the fact that a mural not only modernizes a traditional image carrier, but also intervenes in social space in an emancipatory way, effects subversive impacts, and/or is—as Norma Patricia Lache Bolaños claims without distinction, for all graffiti and street art created in Oaxaca in 2006—an "action of resistance" (2009: 214)?

Subversion and Art

First of all, it should be explained or at least more clearly defined what "subversion" actually means. Subversion, in my attempted definition, consists of practices that challenge a dominant political–moral order in an emancipatory sense and not only sabotage its stability (which is already possible in individual acts), but also undermine it (which generally requires collective efforts) in a long-term perspective. Achieving the long-term goal of subversion (from Latin: *subvertere*, to overthrow or spoil), though, would demand a whole series of subversive acts and has historically never yet occurred solely through these, that is, without political organization or steering. Subversive practices do not necessarily require subversive protagonists and need not be intended as such at all; even unintentional actions and unpremeditated behavior can have subversive effects. But in this attempt at definition, subversion is named on the one hand as a political means and on the

---

photography—following a distinction from Porres Ferreyra (2009)—referred positively to the revolts in their work as well.

6   Some of the works created in the context of the insurrection could be seen beyond Oaxaca, for example in the exhibition "Oaxaca: Aqui No Pasa Nada," held in the Galería de la Raza, San Francisco from October 13 through November 3, 2007 (http://www.galeriadelaraza.org/eng/events/index.php?op=view&id=1008).

other as a means of undermining as opposed to either demanding or attacking. Terrorist (attacking) and labor union (demanding) actions can have subversive effects, but are not themselves subversive.[7] It is of course problematic—and this is probably also the reason why the concept has hardly played a role in the political discourses of the past few years—to assume a relatively stable order sustained by the majority of the population, which is a prerequisite for subversion. Practices can only be subversive from a marginalized position (which, as is generally known, need not always be a minority position); practices of "subversive rule" or "subversive hegemony" are by definition impossible. Even when one considers that subversive practices are not determined generally and metahistorically, but only relatively and concretely—that is, in relation to other practices and institutions—the problematic aspects of assuming stability and consensus remain. Although I have now formulated a general definition, it may still be possible in the case of artistic practices to set up or at least to discuss other criteria, which could form a basis for declaring them subversive.

Criterion 1: Reflecting on Production Conditions, Installing Collectivity

To break through the representation mechanisms of the art field and achieve effects in the field of the political, it may be helpful to first of all reflect in some way on the production conditions of a work of art. The mode of production arranges productions according to field-specific specifications, which must be questioned if effects are to be achieved beyond the field of production. This reflection frequently already results in the partial creation of new modes of production, such as collective ones. Both the reflection and the new modes of production often characterize artistic practicAes that have arisen with social movements.

In Mexico there is a long tradition of artistic practices and formations that have coalesced in the context of social movements. Even the muralism the state promoted in the course of the reconstitution of the nation following the Mexican Revolution (1910–1920) can be regarded as a current of this kind close to the movement. Its leading representatives not only regarded themselves as artists but were also politically active in addition to and, according to their explicit understanding, also through their art: in allusion to the radical peasant leader and revolutionary Emiliano Zapata (1879–1919), Diego Rivera called himself a "Zapatista." Together with the muralists David Alfaro Siqueiros and Xavier Guerrero, Rivera also formed the executive committee of the Revolutionary Union of Technical Workers, Painters and Sculptors, founded in 1922. In 1924 the journal *El Machete*, which later became the official newspaper of the Communist Party

---

7   The concept of subversion employed here is therefore not limited to the dimension of "political-revolutionary subversion," which Ernst et al. (2008: 18–19) distinguish from "artistic-avant garde," minority or underground subversion, and a deconstructivist concept of subversion.

and was named after the tool of the Latin American rural population that is also used as a weapon, developed out of the newsletters of the union. In relation to the status of his person as well as that of his work, however, Rivera remained highly conventional. His fresco "The Making of a Fresco Showing the Building of a City," (San Francisco Art Institute, 1931) depicts the act of collective work, but the individual artist continues to remain literally and metaphorically in the center of the picture as its creator.

The collective Arte Jaguar, which was active in Oaxaca around 2006, ties into this form of representation in the "Making of ..." mold by posting a YouTube film on their MySpace page (http://www.myspace.com/losartejaguar) that shows the creation of a mural in Oaxaca. Yet Arte Jaguar goes far beyond Rivera's reflection on representation by also basing their work on organizational consistency. With their work as a collective, Arte Jaguar belongs more to a tradition that has emerged since the 1920s with and alongside muralism: many of the artists close to the movement already organized their work collectively (as leagues, in unions, movements, or groups) in the immediate aftermath of the Revolution (cf. Híjar Serrano 2007, Audefroy 2009). Collective work can be understood not only as an attempt to counter the structural individualism—the cult of the individual creator, the value of authentic authorship, and the like—of the field of art and its mechanisms of representation oriented to ward individual names with moments of disruption (even though group names can ultimately also be fetishized and made to conform to the market). In some circumstances it can also be opposed in particular to individualistic, sociopolitical models such as neoliberalism. A new phase of collective organizing emerged within the field of art in the 1970s as a reaction to the stagnant representations within the field and in the context of an individualism considered alienated in late capitalist society.

Criterion 2: Implicit References, Explicit Ties

In artistic work signs should emerge or become possible through reflections on the production process that allow to tie in with political and/or social occurrences through purely art-immanent (methodical) references. During the insurrection in Oaxaca, ties were made to the art practices of the 1970s in Mexico in various ways. When Ana Santos applies her shadow figures to the walls of buildings in Oaxaca—an art project she began in 2004—these were initially only anonymous spots, the presence of absent people, sprayed or painted on relatively permanently, although in passing. There have been very similar shadows in Mexico's art history before, when graffiti had only just been invented and most walls were still white in urban space that was barely public, because it was controlled by the state. One of the collectives of the 1970s, which has been registered in art history as the group phenomenon "Los Grupos," attached shadow figures to the walls of fences, stadiums, and other buildings in Mexico City. Grupo Suma, one of the groups of Los Grupos, painted *el burócrato* (the bureaucrat), the silhouette of a man with a briefcase, on the walls of the capital in the late 1970s. What could be read here was a criticism of the bureaucratic rule of the state party PRI. Los Grupos also emerged partly in the context and partly as an effect of social movements, namely those of 1968 (cf. Espinosa 2002, Gallo 2007, García Canclini 2009, Kastner 2009b). Artistic actions like the production of ephemeral wall paintings even accompanied the student movement (cf. Vázquez Mantecón 2007). The transience of the works was to be understood as a direct reaction to the monumental and pedagogical orientation of the post-revolutionary murals. The choice of artistic means thus also reflected the political anti-authoritarian orientation of the student movement. The art movements of the post-1968 era frequently referred, both implicitly and explicitly, to revolutionary Mexican history and operated in this way, for example, through the demand, shared with many social movements, for the realization of the social revolution that had come to a complete standstill in state bureaucracy and the corporative single-party rule.[8] This kind of fight over the revolutionary heritage is evident especially in current social struggles, because the fight over the plausibility of legitimate inheritance always also forms part of the struggle for the legitimacy of current concerns. Thus, forms of symbolic inscription into revolutionary history were not lacking in Oaxaca in 2006 either. The Zapata portrait by Arte Jaguar is always posited in a direct, explicit context of appropriation: the revolutionary hero Emiliano Zapata belongs to the permanent inventory of state revolutionary folklore, adorns Rivera murals as well as T-shirts and other tourist merchandise, and yet still turns up in many contemporary social struggles as an icon as well. Uses of his image or his name in social conflicts also point to the unfulfilled promise of the Revolution from 1910–1920 and stake a claim for a radical political transformation. This is an example of an artistic intervention in a concrete historical–political territory. It aims at questions about the interpretation of the past and its value and valuation for the present. However, it is also possible to intervene in these kinds of value questions without blatant dimensions of political content like the Zapata portraits.

Criterion 3: Targeting Structures of Perception, Engaging in Insurrections

Subversive art is possible if it produces art-immanent—that is, by definition anti-everyday—references, using means and methods (or contents) of art and pointing beyond them at the same time, thus incorporating itself into everyday collective practices without relinquishing its existence as art.

These kinds of everyday cultural practices are currently also increasingly becoming the settings of what Antonio Negri called insurrection.[9] In his model

---

8   On the relationship between art and the politics of remembrance in Latin America, cf. Kastner 2009b.

9   According to Gerald Raunig (2005: 42), the concept first appears up in Negri's essay "Republica constituente," published in Italian in 1993. The moment of constituent

of social struggles, following the diagnosed failure of revolutions taking over state power and guerilla wars, Negri propounds a three-part collective practice consisting of resistance, uprising/insurrection, and constituent power. Gerald Raunig (2005: 40–61) explicitly and extensively points out that this is not a step-by-step model, where one step is taken after another, but rather a three-dimensional, inseparable process. In Raunig's reading, the concept of insurrection relates to modes of subjectivation enabled during collective revolts. Insurrection thus occurs both collectively and individually but is only one aspect of subjectivation: it does not last and, unlike resistance and constituent power, can hardly be organized or steered. "Insurrection is a temporary flare, a rupture, a flash of lightning, in short: the event" (Raunig 2005: 53).

Thus, art is to incorporate itself into a flash of lightning? What could this mean? It could mean the following: the sudden flare of a plausibility (such as that the federal state of Oaxaca is governed in an authoritarian manner and something has to be done against this authoritarianism), the moment of setting off to a practice that belongs to the spheres of the everyday and the non-everyday at the same time (like cooking while guarding the barricades in the city center), the spontaneously thwarting of fixed habits (like actively creating media instead of merely consuming broadcasts). These kinds of everyday practices that simultaneously break with everyday life can be reflected on and/or methodically anticipated through artistic practices: indigenous people carrying burdens, like those sprayed as stencils on public walls by the collective Lapiztola during the uprising, can be incorporations of this kind. On the one hand they reproduce an everyday perception in the urban space of Oaxaca, which is the capital of the Mexican federal state with the second highest proportion of indigenous populations. On the other hand, the everydayness is ruptured specifically through this reproduction (and the manner of reproduction), resulting in a space of imagination that at least potentially enables a vision of everything that is possible, from the concretely shown situation via conditions of working and living to indigenous organizations' demands for land rights. Diego Rivera had already depicted groups of indigenous people, who had hardly been represented in art (and in state politics) before, but in his murals—wholly in keeping with the concept of *Indigenismo*—they served to represent a glorified, nationally functional past rather than functioned as subjects with agency in the present.

Resistance and Negotiation

All three criteria for what is subversive in art have, or rather lay claim to, two interlocking, mutually connected sides: the side internal to the field of art that

power subsequently appears in the books Negri co-authored with Michael Hardt, *Empire* and *Multitude*. Thereafter it became a key concept in post-Operaist-inspired research on social movements (see, for example, Shukaitis, Graeber, and Biddle 2007).

relates to the production context in the narrower sense, and the external side that points beyond the rules of the specific institutions, mechanisms, and practices of art itself. To understand the way in which the two sides interact, one can turn to several theoretical frameworks that also allow for a conceptualization of these kinds of exchange relationships. The approaches of Antonio Negri and Pierre Bourdieu provide two cases in point. Even if it seems audacious to draw on such fundamentally different theoretical frames of reference as the post-Operaist one of Antonio Negri (and Michael Hardt) and the genealogical one of Bourdieu, this "audaciousness" is indeed justified by the reference to contemporary social movements that the two approaches have in common as well as their partial agreement with one another on the role that political resistance has for stability and for the mutability of social conditions.

For Negri and Hardt collective resistance is the precondition for the constitution of social classes and, beyond this, an indication of social conflicts to come. Even though there is certainly nothing like the "primacy of resistance" (Hardt and Negri 2004: 82–84) of post-Operaism in Bourdieu's approach, it seems that the two frameworks share a fundamental emphasis on social struggles. However, these struggles play a central role for Bourdieu that is not limited to field-internal confrontations. Indeed, he states in reference to the entire social space: "History only exists as long as people rebel, resist, react" (2006: 133). This kind of resistance, understood as the active questioning of the given as defined by minimum consensus—an originary act for Hardt and Negri and a reactive one for Bourdieu—, in both theoretical frameworks means establishing politics as a sphere of contentious social order. The question of the subversion of art always implicitly aims at the contribution of art to this establishment of politics. The answer to this question generally does not merely seek to undermine the art system and its field mechanisms, but instead applies them to the aforementioned, existing political–moral order. What should be emphasized at this point (and also highlighted as a further partial agreement between the approaches of Negri and Bourdieu) is that this order does not consist solely of state institutions and their legislative, juridical, and executive branches. For Negri and Hardt, politics means, among other things, the capability of collective actors and institutions "of entering into societal conflicts and differences and negotiating them" (Negri and Hardt 1997: 93). These kinds of conflicts and differences are tied to modes of social production and reproduction; their processes of negotiation do not begin in institutional bargaining. In reference to the state form of neoliberalism, Negri and Hardt note that the state—contrary to all ideological definitions of its withdrawal—is certainly capable of founding a moral unity and consensus (1997: 101). They start from this foundation of consensus in particular in order to be able to show that even the neoliberal, postmodern state does not lose its power, although it increasingly refuses the official procedure of negotiation. Pierre Bourdieu described this level of politics, which aims at founding consensus and at standardization, at length. It is based on the kinds of negotiation processes that do not first take place in representative and institutional procedures but already in the battle over structures of collective thinking and perception. In Bourdieu's terminology these can be grasped as the symbolic dimension of the political. And these structures of thought, emotion, and perception are also that through which artistic practices can have an impact that is political as a whole and ultimately specifically subversive.

The Symbolic and Politics

If one raises the question of the effects of artistic practices that reach beyond the art field, that protrude into the political field, then it must be recalled that the political field encompasses far more than political parties, parliament, and the other state apparatuses. Bourdieu's field concept was developed, after all, specifically as distinct from Louis Althusser's term of the "ideological state apparatus" to overcome the notion of a relatively closed "apparatus" occupied by bureaucratically organized personnel and oriented to a certain purpose.[10] With a notion of the political as a terrain of the struggle over structures of perception and thought, Bourdieu is also able to describe political conflicts as a theoretical and practical fight for power, in which the goal is to "assert the legitimate view of the social world" (Bourdieu 2001: 238). More precisely, political conflict always involves "recognition accumulated in the form of symbolic capital in reputation and respectability, which empowers determining legitimate knowledge and a sense of the social world, its current significance and the direction in which it will and should develop" (2001: 238).

At the same time, the symbolic dimension of the political must also be defended against two relative conflations of Bourdieu's theory. In her engagement with the political field of Mexico between 1968 and 2000, Martha Zapata Galindo describes, against the background of Bourdieu's theory, "political power" as personally transferable power over means of production and reproduction. She differentiates this from "scientific power," which can hardly be passed on, because it is based on prestige tied to a person (cf. Zapata Galindo 2006: 59). In this perspective, politics or the political means nothing other than the state apparatuses and their actors. Even though Zapata Galindo's concrete analysis of the political role of intellectuals in Mexico in the 1970s is pertinent (and also fascinating with the detailed data material), the form of the political conceptualized in this way still contains a constraint that, as shown above, can hardly be reconciled with the idea of the field concept. According to Zapata Galindo, artistic practices, which she purposely omits

---

10   In direct reference to Althusser, Bourdieu writes: "In a field there are struggles, thus history. ... The school system, the state, the church, political parties or unions are not apparatuses but fields. In a field agents and institutions struggle, with different degrees of power and thus prospects of success, according to the regularities and rules constitutive of this space of play (and in certain given conjunctures, over the rules themselves), over the appropriation of the specific profits that are in play in this game" (2006: 133).

in her analysis that concentrates on the literary field, could not be politics in their collectivity and in their actions in public space, but can only be art.[11]

In a similar vein, Isabell Graw (2008) ignores the link Bourdieu constructed between the symbolic and the political in her analysis of the field of contemporary art inspired by Bourdieu, when she relates, in a kind of mirror inversion of Zapata Galindo's contraction, the symbolic value of artistic practices only to their commodity value[12] but not to values in the ethical–moral sense and as a set of embodied standards. For it is precisely their significance as unconsciously guiding practice that Bourdieu addresses: understood as legitimate knowledge about and sense of the social world, the symbolic becomes a key category of political conflict. On the level of the symbolic, collective social power relationships are consolidated in individual as well as in collective corporeal dispositions. These mutual (or even contradictory) attitudes are the basis for all measures aiming at the conflictual regulation of that which is shared—in other words, at that which is to be understood as the political.[13]

If the political has now been conceived to the extent of including the struggle over the significance of the social world, two things become possible: one the one hand, we can measure the political content of those artistic actions that emerge from the cultural field to join this struggle and thereby also temporarily and partially overcome the limits of the field. Even as the commodification of art increases, in the process of which the work of art with its functions as a source of belonging and

---

11   Zapata Galindo follows the thesis formulated by César Espinosa (2002) and others that the gradual demise of the hegemony of the state party PRI began when the student movement was crushed on October 2, 1968, and she affirms this particularly in reference to the cultural field.

12   In symbolic value, according to Isabelle Graw, the "special status of art, historically much fought for" (2008: 32, translation mine) finally found itself. It incorporates the demands on art that have been more strongly formulated since the eighteenth century; in symbolic value "that symbolic meaning, difficult to pin down, which is composed of various factors—singularity, art-historical description, establishment of the artist, promise of originality, assurance of duration, postulation of autonomy or intellectual demands" (2008: 32, translation mine).

13   It is only on this level of symbolic relationship that we can also answer, for example, the question of why the classes who are being ruled uphold the ruling class and actively contribute to its organization. For they do this not necessarily and primarily because of "conscious and considered agreement," but rather under the influence of the power "that in the form of perception schemas and dispositions … has in the long run enlisted the bodies of the ruled" (Bourdieu 2001: 219, translation mine). The symbolic relationships are imposed according to Bourdieu "on the subjects as a system of rules that possess absolute validity within their domain as a system that cannot be reduced either to the rules of play in the economic sector or the particular intentions of the subjects" (1974: 73, translation mine). With regard to social transformations one should accordingly address not only the economic relationships or the appeal to the cognitive "intentions of the subjects" but also symbolic relationships and cultural leadership that have settled into the bodies of the ruled in the form of perception schemas and dispositions.

a guarantor of distinctions is more and more to be seen as "the forerunner of the brand-name article" (Graw 2008: 136), the artistic work is not a product like any other. Neither is its symbolic value completely absorbed by its commodity value, nor does it remain limited to the battles over differentiation that are immanent to the field. If artistic work is always work on the symbolic, this work can under certain conditions bring about greater or lesser effects in the social realm.

If we consider this symbolic dimension of the political, the discussion of the politics of artistic actions can be move beyond concrete examples such as the uprising in Oaxaca. For struggles over symbolic capital and legitimate knowledge about the sense of the social world occur in all societies and under the most different political administrative systems.

Insurrectional Flashes of Lightning and Symbolic Labor

The art works created on the exterior walls of buildings in Oaxaca during the insurrectional flashes of lightning can be designated as part of a certain symbolic work in Bourdieu's sense. This is the symbolic work required, according to Bourdieu (2001: 241), "to elude the mute evidence of the *doxa* and to articulate and denounce the arbitrariness it veils." *Doxa* is the term Bourdieu uses for self-evident knowledge or, more precisely, the schemata of perception, on which perspectives and ways of thinking are founded.[14] By developing symbolic forms that are integrated into the political struggles of various artistic aesthetic traditions, the works discussed here merge into the work against the "veiled arbitrariness" of the *doxa*. On the one hand they fulfill the classical enlightenment services of uncovering and denouncing. As symbolic forms, on the other hand, they at the same time reach levels that are more rarely touched by conventional political activism aimed at contents free from ambivalence. This is not intended to maintain that artistic practices generally go deeper or achieve more directly interventionist effects in society. Yet the field-specific professional way of dealing with the shift of the meaning of signs—picturing the absent as well as taking up, appropriating, repeating, and shifting this kind of work on symbols has belonged to the "business as usual" of artistic practices since the beginning of modernism—contain at least a certain potential in terms of the symbolic in social and political conditions. In this way, artistic practices can also influence the incorporated system of rules that is to be understood as the political across different fields.

By devoting themselves to these inscribed rules that have become everyday life, artistic practices also place themselves in the context of a reappropriation of the

---

14   According to Bourdieu, determining and shaping these foundational schemata of perception is claimed by the social forces that condense in the state. An exploration of a possible link between Bourdieu's approach and more recent materialist state theory in the following of Nicos Poulantzas (cf., for example, Bretthauer et al. 2006) through this question of the condensation of societal force relations is unfortunately still a desideratum.

political: for, first of all, the actions described affect the "veiled arbitrariness" and the "mute evidence" of the everyday perception of the political. The insurrection effects or enables questioning the political at a given moment, from within everyday life—in other words without prior sociological analysis or reflection from a political science perspective. The artists active during the uprising support this questioning with their concrete artistic activity, because in a certain sense they exemplify this question themselves as well as in and with their own practice: painting on the street not only exposes the artists' own work to direct observation by passers-by (thus disrupting the artistic norm of the individual creative process in sheltered retreat) but also confronts these viewers with the art historical tradition of muralism and with the break with this tradition at the same time. This demonstrates that political murals can be painted not only in the service of the national project and the state party but also explicitly against the political administrative establishment. As this embeddedness in previous practices in the artists' own field of production shows, working on the symbolic is never without preconditions, but it exists in a permanent process of appropriating and being appropriated. "It is only when the inheritance has appropriated the inheritors," writes Daniel Bensaïd, conveying a central idea from Bourdieu's habitus concept, "that the inheritor can procure the inheritance" (2006: 105–6). Allowing oneself to be seized by the inheritance—that is, an implicit reflection on one's own conditions of production and reproduction— also seems to be an important precondition for the success of artistic interventions. Yet these strategies also work as a reappropriation of the political to the extent that they operate, secondly, to counter attempts to close the political field. As they are clearly recognizably in the context of political mobilizations represented by the APPO, they reclaim the political that is claimed by the professionals for everyone, thus taking a stance for a *political perception of the everyday*. By making the everyday a relevant component of political confrontation, the artist groups, similar to social movements, "illegally exercise politics" (Bensaïd 2006: 109), which professional politics takes action against (and which Zapata Galindo does not take into consideration as such).

The illegal exercise of politics is subversive. Art practices like graffiti and street art can primarily develop these effects when they are able to maintain— to summarize my thesis—the interplay of movement between the everyday and the non-everyday as described here, when they are able to temporarily establish themselves in this shift back and forth between graffiti/street art as art, on the one hand, and as part of the practices of social movements, on the other.[15] The hybrid form or this maneuvering back and forth hinders or blocks, first of all, their hasty functionalization as a (purportedly functionless and hence "artistic") object of prestige and the subsequent pure commodification. Not every legitimization

---

15   The acceptance of graffiti and street art as art is not necessarily accompanied—as Baeumer (2009: 114) maintains—by ignoring the praxis of the production process that belong to them. Indeed, an inclusion of graffiti/street art in the praxis of social movements instead counters this omission.

of artworks automatically leads to their commodification. In the context of struggles over the city, recognition (as art) can even serve effects external to art. Secondly, this hybrid constitution makes it more difficult for the murals to devolve to the vacuous level of significance of spots of color or crumbling plaster that characterizes the urban building facades no more and no less than the forms that claim something different in form (and content). The latter is also the reason why, in terms of effectiveness, symbolic work on their—at least temporary— elevation as art practice is to be adhered to; an adherence, on which the distinction between everyday objects and art objects (which theoretically seems antiquated and is politically often disparaged), is ultimately also based. Although there is no direct, proportional relationship between the appreciation and impact of cultural production, without recognition, that is, without being furnished with symbolic capital, there is a tremendous drop in the potential for effects within the field of production and even more so outside it.

Consequently, not every piece of graffiti implies an act of taking possession of public space, as García Canclini maintained. Graffiti functions in this sense especially when it can establish itself as a specific artistic practice and make this specificity appear irrelevant at certain moments by merging into other cultural practices. This works best in those moments in which the everyday takes on non-everyday forms and in which other non-everyday manifestations outside the art field belong to everyday life. In short: in situations that are insurrections.

Translated from German by Aileen Derieg.

Works Cited

Audefroy, J. 2009. Estética socialista y cuidad. *Cuidades. Análisis de coyuntura, teoría e historia urbana*, 83, 8–18.

Baeumer, T. 2009. Zeichen setzen! P.S. Grafitti sind Krieg, in *Kommt herunter, reiht euch ein ...: Eine kleine Geschichte der Protestformen sozialer Bewegungen*, edited by K. Schönberger and O. Suter. Berlin and Hamburg: Verlag Assoziation A, 109–29.

Beas Torres, C. 2006. El fantasma de Louisa Michel. *La Jornada*, September 30 [Online]. Available at: http://www.jornada.unam.mx/2006/09/30/021a2pol. php [accessed: 5 July 2010].

Bensaïd, D. 2006. *Eine Welt zu verändern: Bewegungen und Strategien*. Münster: Unrast.

Bourdieu, P. 1974. *Zur Soziologie der symbolischen Formen*. Frankfurt/M.: Suhrkamp.

Bourdieu, P. 2001. *Meditationen. Zur Kritik der scholastischen Vernunft*. Frankfurt/ M.: Suhrkamp.

Bourdieu, P. 2006. Die Ziele der reflexiven Soziologie: Chicago-Seminar, Winter 1987, in *Reflexive Anthropologie*, by P. Bourdieu and L. Wacquant. Frankfurt/ M.: Suhrkamp, 95–249.

Bretthauer, L. et al. (eds.) 2006. *Poulantzas lesen: Zur Aktualität marxistischer Staatstheorie.* Hamburg: VSA.

Collective Reinventions. 2009. Barrikaden: Der Aufstand von Oaxaca (Mexiko): Sein Sieg, seine Niederlagen und darüber hinaus. *Kosmoprolet,* 2, 132–63.

Ernst, T. et al. 2008. SUBversionen: eine Einführung, in *Subversionen: Zum Verhältnis von Politik und Ästhetik in der Gegenwart,* edited by T. Ernst et al. Bielefeld: Transcript, 9–23.

Espinosa, C. 2002. Introducción: bifurcaciones: l. Los años 70 y los años 90, in *La Perra Brava: Arte, crisis y políticas culturales,* edited by C. Espinosa and A. Zuñiga. México, D.F.: UNAM, 17–42.

Gallo, R. 2007. The Mexican Pentagon: adventures in collectivism during the 1970s, in *Collectivism after Modernism: The Art of Social Imagination after 1945,* edited by B. Stimson and G. Sholette. Minneapolis and London: University of Minnesota Press, 165–90.

García Canclini, N. 2005. *Hybrid Cultures: Strategies for Entering and Leaving Modernity.* 2nd Edition. Minneapolis and London: University of Minnesota Press.

García Canclini, N. 2009. Interview mit Samuel Morales und Mariana Huerta Lledias, in *Zwischenzonen: La Colección Jumex,* edited by Fundación/Colección Jumex and MUMOK: Museum Moderner Kunst Stiftung Ludwig Wien. Ecatepec de Morelos and Wien: Fundación Jumex–MUMOK, 265–77.

Graw, I. 2008. *Der große Preis: Kunst zwischen Markt und Celebrity Kultur.* Köln: DuMont.

Hardt, M. and Negri, A. 2004. *Multitude: Krieg und Demokratie im Empire.* Frankfurt/M.: Campus.

Hepp, A. 2009. Néstor García Canclini: Hybridisierung, Deterritorialisierung und "cultural citizenship," in *Schlüsselwerke der Cultural Studies,* edited by A. Hepp et al. Wiesbaden: VS Verlag für Sozialwissenschaften, 165–75.

Híjar Serrano, A. (ed.) 2007. *Frentes, coaliciones y talleres: Grupos visuales en Méxixo en el siglo XX.* México, D.F.: Casa Juan Pablos et al.

Kastner, J. 2009a. Zeitgenössische Kunst als erinnerungspolitisches Medium in Lateinamerika. *Atención! Jahrbuch des Österreichischen Lateinamerika-Instituts,* 12, 191–213.

Kastner, J. 2009b: Praktiken der Diskrepanz:. Die KünstlerInnenkollektive Los Grupos im Mexiko der 1970er Jahre und ihre Angriffe auf die symbolische Ordnung, *Atención! Jahrbuch des Österreichischen Lateinamerika-Instituts,* 13, 65–80.

Lache Bolaños, N.P. 2009. La calle es nuestra: intervenciones plásticas en el entorno de la Asamblea Popular de los Pueblos de Oaxaca, in: *La APPO ¿Rebelión o movimiento social? (Nuevas formas de expresión ante la crisis),* edited by V.R. Martínez Vásquez. Oaxaca: Cuerpo Académico de Estudios Politicos, 199–217.

Löw, M. 2001. *Raumsoziologie.* Frankfurt/M.: Suhrkamp.

Löw, M. 2008. *Soziologie der Städte.* Frankfurt/M.: Suhrkamp.

Martínez Vásquez, V.R. (ed.) 2009. *La APPO ¿Rebelión o movimiento social? (Nuevas formas de expresión ante la crisis)*. Oaxaca. Cuerpo Académico de Estudios Politicos.

Negri, A. and Hardt, M. 1997. *Die Arbeit des Dionysos: Materialistische Staatskritik in der Postmoderne*. Berlin and Amsterdam: Edition ID-Archiv.

Nevaer, L. 2009. *Protest Graffiti Mexico Oaxaca*. New York City: Mark Batty.

Osorno, D.E. 2007. *Oaxaca sitiada: La primera insurrección del siglo XXI*. México, D.F.: Grijalbo/Random House Mondadori.

Porras Ferreyra, J. 2009. Las expressiones artísticas y la partizipación política: el conflicto oaxaceño de 2006, in *La APPO ¿Rebelión o movimiento social? (Nuevas formas de expresión ante la crisis)*, edited by V.R. Martínez Vásquez. Oaxaca: Cuerpo Académico de Estudios Politicos, 219–45.

Raunig, G. 2005: *Kunst und Revolution: Künstlerischer Aktivismus im langen 20. Jahrhundert*. Wien: Turia und Kant.

Reinecke, J. 2007. *Street-Art: Eine Subkultur zwischen Kunst und Kommerz*, Bielefeld: Transcript.

Shukaitis, S., Graeber, D., and Biddle, E. 2007. *Constituent Imagination: Militant Investigations / Collective Theorization*. Edinburgh and Oakland: AK Press.

Vázquez Mantecón, Á. 2007. La visualidad del 68, in *La Era de la Discrepancia: Arte y cultura en México 1968–1997*, edited by O. Debroise. México, D.F.: UBAM—Turner, 34–36.

Zapata Galindo, M. 2006. *Der Preis der Macht: Intellektuelle und Demokratisierungsprozesse in Mexiko 1968–2000*. Berlin: Tranvia/Walter Frey.

Zires, M. 2008. Estrategias de communicación y acción política: movimiento social de la APPO 2006, in *La APPO ¿Rebelión o movimiento social? (Nuevas formas de expresión ante la crisis)*, edited by V.R. Martínez Vásquez. Oaxaca: Cuerpo Académico de Estudios Politicos, 161–197.

Chapter 4

Black Day in the White City: Racism and Violence in Sucre

Juliana Ströbele-Gregor

"*Sucre—la ciudad blanca.*" The colonial city of Sucre, located in the heart of Bolivia and its constitutional capital, takes pride in its name. It comes from the city's whitewashed baroque churches, monasteries, colonial mansions and residential neighborhoods as well as from its neatly arranged squares and streets. Sucre, a UNESCO World Heritage Site.

The name of Sucre recalls the colonial cultural heritage of one of the most beautiful colonial cities in Latin America, having been thus marketed for tourism. However, with the (re-)construction of the city's urban space, the conservation of its colonial and postcolonial architecture, and the reinvigoration of its sites of cultural interest, baroque musical tradition, and historical narrative, a *social identity* based on the colonial cultural heritage is being constructed, as well. In the "white city," the ideology of "whiteness" of the *criollos*, the dominant population group, is always present. This ideology combines the intensive cultivation of the colonial past with the self-perception of the middle and upper classes as *criollos* or *criollos-mestizos*, both as the descendents of the Spanish colonial rulers and aristocrats and as the freedom fighters of Spanish descent in the war of independence in the beginning of the nineteenth century. In the identity construct of the *criollos* in Sucre, the following elements are proudly merged together: colonial cultural heritage, identification with the struggle for independence, local history within the context of the founding of Bolivia, and remembrance of the city's powerful silver oligarchs, *hacienda* owners, military leaders, intellectuals, and artists. According to this self-perception of the urban space in "their" city of Sucre, however, the indigenous population is relegated to the periphery, outside of the "civilized urban society" (Defensor 2009: 27). In the socio-racial stratification between whites and Indians (Postero 2008: 28) that characterizes society in Sucre, unwritten and historically-developed lines of demarcation and norms give its urban space its structure. They mark where "*indios* belong"—and where they do not. Paradoxically, the everyday reality of Sucre, in which the lifestyles of the middle and upper classes would be unthinkable without the services provided by and the diverse economic activities of the indigenous and *mestizo* population, is not perceived as a contradiction within the exclusive self-awareness of *criollos* as such. Regarding the construct of the image of the city, however, the indigenous population is made invisible. The fact that "*indios*"—a derogative terms for the indigenous population—are part of the marketed culture of Sucre does not change this exclusionary invisibility, for here they are folklorized and construed as sort of "living memories of the indigenous tradition" and, as such, are thus offered as tourist attractions. The best example of this concerns the *campesinos,* who appear in their vibrantly-colored clothes at the Sunday market in the small city of Tarabuco. This motif can be found on every postcard of and advertisement for Sucre; excursions to the market in nearby Tarabuco are seemingly mandatory activities for all visitors to the city. From the perspective of the *criollos,* the indigenous communities in the region surrounding Sucre represent the *"indios,"* the "others," in a distinct rural world in which, as is commonly stated, time has stood still. In the construct of the colonial cultural heritage, these "others" do not belong in the city.

Due to the fact that this "white" history of Sucre is so clearly cultivated in the imagination of the *criollos*—especially as such reasoning serves as a covert legitimization of social exclusion and ethno-social hierarchy—it is important to succinctly address a few points regarding the history of Sucre.

In 1538, only a few years after the *conquest* of Peru under Francisco Pizarro in 1531, the town of Villa de Chuquisaca was founded in the region south of Lake Titicaca, which the Spaniards named Charcas. Shortly thereafter, it was renamed La Plata. After independence, it was renamed again in honor of the revolutionary military leader Antonio José de Sucre. More than any other city in the southern Andes region, this city reflects the history of Spanish rule and the founding of Bolivia after independence in 1825. Not far from the silver mines of Potosí, located in a fertile valley region, thus serving both as a trade hub between the Peruvian region and Rio de la Plata with its access to the Atlantic, and also acting as the seat of the *Audiencia de Charcas*; La Plata was one of the political and cultural centers of the Spanish colonial empire in South America. The silver deposits and trade made La Plata wealthy and allowed the Spaniards to live in luxury—at the expense of the indigenous population who served as forced labor in the mines and *haciendas.*

In 1625, a university was founded, and its reputation soon attracted scholars and students from many regions of the subcontinent. The university continues to be a great source of pride for the *criollos,* as Sucre has remained a point of convergence for scholarship. Today, its students come from every region of the country, especially from the departments of the lowland region, the *Media Luna,* where the centers of the militant opposition to the government of Evo Morales are located. As the construct of a "white identity" is cultivated in these regions, the university students and young people who originated there were surely participants in the growing social and political tensions which led to the violence on May 24, 2008 discussed in this article.

The region of Charcas was one of the starting points of the anti-colonial struggle for independence in the southern Andes region. The first revolts of the descendents of the Spanish against colonial rule in today's Bolivia took place in Sucre in 1809—the famed *primer grito por la libertad* ("first cry for freedom").

Thus began the 16-year-long war of independence culminating in the founding of the state of Bolivia in 1825.

In contrast to the rebellion of the Aymara under Túpac Katari against the Spaniards in 1781, which aimed at attaining sovereignty of the southern Andean peoples, the rights of these indigenous peoples hardly played a role in the war of independence in Bolivia. Even though indigenous people served in the war, their hopes for autonomy were shattered shortly thereafter. The goals of abolishing oppression and exploitation fell quickly by the wayside, suppressed by the economic interests of the new state and *criollo* oligarchy. Civil rights remained limited. The postcolonial society continued to be dominated by extremely vertical socio-racial structures and the *de facto* political exclusion of the indigenous rural population. In this sense, even today, many inhabitants of Sucre misuse the festivities commemorating the *primer grito por la libertad* as a reminder of the desire for independence on the part of the *criollos*.

Today in Sucre, as in other regions of Bolivia, a separation from the "*indios*" continues to shape the habitus of those social groups that construct themselves in terms of "whiteness," even though the acknowledgement of this racist attitude is not accepted as correct in the liberal political discourses of public sphere. Within the context of the festivities commemorating the *primer grito por la libertad* on May 24, 2008, however, this racism became clearly evident.

Criollos against Indios: The Black Day

"De rodillas indios de mierda, griten viva la capitalidad!" "Sucre se respeta carajo!" "Llamas, pidan disculpas!" ("On you knees, bloody indios, go on, say it: long live the capital status of Sucre!" "You have to respect Sucre, carajo!" "You lamas, ask for forgiveness!"). These were the slogans used by the adolescents of the Chuquisaca department, mostly comprised of students as well as activists of the Departmental Autonomy Movement, when they attacked the indigenous representatives of village communities at the city gates of Sucre on the May 24, 2008 (Contreras 2008). The indigenous local council members, mayors, men, women, and children, all Quechua campesinos, had traveled into town for the Independence celebrations. They were planning to show their respect and gratitude to president Evo Morales for the tractors and investments supplied to their region. Eye witnesses, including international scientists (Platt, Martínez, and Riviere 2008), reported in various media (see, for example, ComunicaBolivia 2008) that the campesinos were rudely insulted, kicked, and beaten—even women and children. Fifty-five campesinos, mostly indigenous dignitaries, were taken as hostages. They were taken to the main square, in front of the venerable building *Casa de la Libertad* (Freedom House), where the city authorities were holding the celebrations. Amongst insults and hoots of the crowd, the young people forced the campesinos to circle the square, strip down to their shirts, kneel down, and ask the inhabitants of Sucre for forgiveness for the violent struggles during the finalization of the Constitutional

Assembly (*Asamblea Constituyente*) in November 2007. Forced to kiss the ground and set fire to traditional ponchos and insignia, the campesinos were also forced to burn the Whipala, the flag of the indigenous people of the Andes, and also the flag of the governing party *Movimiento al Socialismo* (MAS). Further humiliations included being forced to salute the flag of Sucre and repeat slogans against the government of Evo Morales and for departmental autonomy. In other parts of the city, inhabitants of Sucre—most often drunk—abducted campesinos into houses, where they kept and abused them for several hours. Reports of captives mention beatings and other abuse. Some were even forced to eat chicken excrements. There have also been reports of women being raped (Platt et al. 2008). Official numbers record more than fifty wounded, several severely (other sources speak of thirty-five [Contreras 2008]). Many of these victims were denied medical help by the two main hospitals (Santa Barbara and the university hospital). Identity papers, watches, and what little money victims carried with them were all stolen. The president suspended his trip to the celebrations and demanded judicial prosecution and strict condemnation of the intellectual originators of the violence as well as the culprits. The Lord Mayor of Sucre and the Interinstitutional Committee were quick to apologize, but it was their political discourse that had sparked the conflagration. This paradox leads to the question that I will try to answer in the following essay:

What is expressed in the gesture of the half naked, kneeling indigenous representatives and the insults that were forced upon them? For a better approximation of the events, I will offer a contextualization of the political arena in which they happened as well as an in-depth reflection on the conflict configuration.

The Political Context and Background of the Conflict between the *Media Luna* and the Government[1]

The "Black Day of Sucre" is embedded in a political confrontation that reached a new dimension in the uprising of 2003—and since the 2005 election of Evo Morales as president has revealed the quasi-division of the country. In October 2003 a local confrontation between indigenous villagers and the military escalated into an uprising against the government in the cities of El Alto and La Paz. The protagonists were a diverse mix of social movements, anti-capitalist pressure groups, and, in particular, the indigenous-mestizo population. They protested against the selling-out of the country in the context of privatizations, poverty, and inequality. A fundamental demand of the "Agenda of October 2003" was the call for a "people's assembly" to work out a new constitution, one where indigenous people would be granted equal participation in its drafting. These protests and

---

1   For this chapter I return to previous publications, especially Ströbele-Gregor 2007, 2008a.

demands should be seen as a result of social experiences within a state, with which a large part of the populace could no longer identify, as well as their combined determination to alter society.

Bolivia has a deeply divided society and the roots of this division reach far back into history. Conflict lines run between poor and rich, dividing the small, wealthy, and Western-oriented upper classes in the cities and the indigenous or indigenous-*mestizo* immigrants from the rural, poverty-stricken areas as well as the indigenous rural population living in wretched conditions. They run between the small political and economical power groups who have been able to manage the state to suit their own personal benefit and the indigenous-mestizo population. Until Evo Morales became president, the indigenous population, in particular, had been largely excluded from nearly all political and economic decisions. Morales won the elections with a resounding fifty-three per cent approval, notably because of the votes of the indigenous rural population and the indigenous-*mestizo* urban classes who demanded participation in state matters and a sharing of power. These different grounds for dissention all come together within regional polarizations, which are expressed in the current conflicts, and which run among the Andean highlands and the departments of the *Media Luna*: the eastern, southern, and northern departments of the lowlands.

Here, the social and economic extremes are particularly distinctive: the resource-rich and economically prosperous *Media Luna* is controlled by urban, Western-oriented middle and upper classes of European descent. They align themselves with Brazil rather than with the populous, economically impoverished Andean highland departments such as La Paz, Oruro, Potosí and the parts of Cochabamba where the so-called "*indios*" live. Chuquisaca, in which the formal capital Sucre is situated, has attached itself politically to the *Media Luna*. Although Chuquisaca no longer belongs to the rich departments, to Sucre's middle and upper classes, in particular, the colonial inheritance and the memory of the former importance of the departmental capital city is still very current. Sucre was the city of the silver oligarchs, the colonial "nobility," who maintained their social position even in early republican times. Sucre was the first capital of the republic and seat of government. With the recession of silver mining and the emergence of the tin economy on the Altiplano, the economic and political center shifted toward La Paz, the current seat of government. The silver oligarchs lost their influence during the civil war of 1899, in which the conservatives of Chuquisaca fought against the aspiring economic classes of La Paz (the Liberals) for predominance. The fact that Sucre had to relinquish the status of seat of government to La Paz (Klein 1984: 204–5) remains an open wound to this day.

The conflict currently causing such uproar between the *Media Luna*, in particular the Citizen Committee (*Comité Cívico)* of Santa Cruz, and the central government, started a long time before "*indio*" Evo Morales was elected as president in 2005. The initial catalyst of the 2003 uprising was a decision of the government concerning the energy sector. Public outcry was caused by detrimental contracts with international natural gas companies, which called for the building of a pipeline solely for export purposes and the far-too-cheap sale of the last natural resources whilst the population was not supplied with sufficient amounts of fuel and gas even though Bolivia's available natural resources could meet their demands. Particularly disadvantageous to the lower classes and the rural population, this neoliberal economic policy deepened the gap between the rich and the poor. The fundamental demands of the insurgents were: changes to the economic policy that would take the interests and needs of the poorer classes much more strongly into account, the nationalization of the gas and oil sectors, and the resignation of the government (Piepenstock, Vargas, and Goedeking 2004). The next government under Carlos D. Mesa (2003–2005) hastened to carry out a referendum to resolve the natural gas issue. As a result, the contracts with the international companies were renegotiated and a restructuring—but not the nationalization—of the gas sector was decided upon (Piepenstock, Vargas, and Goedeking 2004). This led to confrontations with the economical power groups of the *Media Luna* who had benefited from the neoliberal economic policies of previous governments; they were unwilling to share their previous profits from the natural gas trade. They were even less willing to approve nationalization measures as demanded by Evo Morales' party MAS and made threats of secession from the Bolivian State. With the assumption of office by Morales, several points of conflict came to the agenda: for example, the nationalization of the gas and oil sectors and a new economic policy based on the requirements of the people. Further points were the drafting of a new constitution which has been one of the demands of the indigenous movement since the 1990s. For Morales and his followers, the implementation of this demand has been the principle item of the political change, *el cambio*. Another point of conflict concerns a long-standing demand of the rural population, their allied organizations, and small leftist parties: the reform of the agrarian legislation and the planned abolition of large land holdings.

In 2006, the Constituent Assembly was elected. Using systematic obstructionist policies, right-wing parties tried to prevent the constitutional bill from passing. However, the bill passed the assembly on December 7, 2007 and was dismissed as illegal by the government opponents in the resource-rich lowland departments Santa Cruz, Pando, Beni, Tarija, and Chuquisaca. Their dismissal is somewhat justified, as the passing of the bill occurred via an extremely dubious process, using only the votes of the governing party, MAS, and her allies; the right-wing parties refused to participate in the voting process.

The confrontation became more heated with the government's intention to reduce the shares of the departments (and the universities) in the gas trade, in favor of financing pensions and other social benefits. The new agrarian legislation in the constitutional bill also met with substantial resistance from the right-wing parties and lowland departments. They rejected the abolition of large land holdings and the transfer of unproductively-used land to the landless—*campesinos* and indigenous groups. This would have particularly affected the *latifundios* in the *Media Luna*, where a few families own enormous land holdings. The main trades here include the breeding of livestock and agricultural business, and enormous areas of land are often claimed for speculation.

In response to the policy of the *cambio*, which is connected to a strong centralization of the state, the demands for autonomy grew stronger in the *Media Luna*. Sustained by the citizen committees, the "autonomy movement" has grown stronger during the last two years in these departments. During the autonomy referendum of 2006, which was carried out in the departments, they had declared themselves in favor of autonomy. They therefore felt legitimately able to formulate their own autonomous bylaws (*Estatuto Autonómico*). The National Election Court as well as the Organization of the American States (OAS) declared the vote regarding individual autonomous bylaws illegal, as the issue needed to be settled in the forthcoming constitution. Despite this ruling, in May 2008, Santa Cruz put its "autonomous bylaw" to the vote and the departments of Pando, Beni, and Tarija followed. In all these departments, the bylaws gained huge approval. In Chuquisaca, the "autonomy movement" lost in the national referendum of 2006. This defeat did not stop radical activists from Sucre from further trying to mobilize people for the movement. They should be considered accountable for the Black Day of Sucre on May 24, 2008.

The bylaws—those of Santa Cruz in particular—represent not only a challenge to the government, but even border on secession.[2] If one examines the logics of the laws, the authors' mindset becomes obvious: of the twenty-eight members of the departmental board that would be leading the government, there are only five seats allotted for indigenous people, completely marginalizing them as a political power, even though they comprise thirty-seven per cent of the department Santa Cruz's population, according to the 2001 census (Instituto Nacional de Estadística de Bolivia 2001). This discrepancy again underlines the division between the "whites" and "indigenous" populations (Ströbele-Gregor 2009).

The self-appointed elite of Sucre has close relations with the Citizen Committee of Santa Cruz. They are joined in opposition to the Morales government. Confrontations occur in several fields. A lynchpin in the conflict is the argument about the capital issue. Fuelled by the political Right, notably the Citizen Committee of Santa Cruz, regional pride in the old capital Sucre was methodically fuelled. The political authorities of Sucre demanded the city's reinstatement as the seat of government and a vote to that effect in the Constituent Assembly. If their demands were not met, then the Constitutional Assembly, which assembled by law in Sucre, would be interrupted. Their slogan was "*capitalía plena*." In a clumsy political move, the ruling party declined to put the capital issue on the agenda of the Constituent Assembly. In mid-August 2007 the convention was promptly interrupted. The conflict erupted into the streets. Violent youths faced off against determined indigenous peasant organizations. Over the course of several months,

---

2   The authority of the central government is reduced to zero. Tax revenue from the gas and oil trade stay within the department. Government agencies are disbanded and replaced by departmental ones. Large land holdings remain unchanged.

the opponents of the constitution used violent riots and road blocks to hinder the convention. In the thick of it was the *Unión Juvenil Cruceñista,* the thuggish youth wing of the *Comité Civico* of Santa Cruz. Human Rights organizations blame them for the worst atrocities. During the tumults in November 2007, three people died. The Interinstitutional Committee has blamed the deaths on the MAS supporters and the minister of the executive office. So far the culprits have not been identified.

Since the confrontations concerning the capital issue, there have been racist attacks not only in Sucre but also in several other places in the *Media Luna.* Offices of indigenous organizations have been torched, and students have assaulted the peasant market in Tarija, shouting "*campesinos, indios* get out of town." Pictures of attacks on women in indigenous peasant clothing were published in newspapers. Racist rhetoric has become common place in discussions of the upper and middle class and not only in the *Media Luna.* It should not been ignored that, on the other side, radical indigenous activists and members of indigenous social movements have made racist comments against "the whites." This particularly concerns the radical indigenous supporters such as the party *Movimiento Indígena Pachakutik* (MIP) (Albó 2002: 86–8, Goedeking 2002, Ströbele-Gregor 2006). Meanwhile, the government team is building on "interculturality."

This is the political context in which the dynamic of the recent violent confrontations during the political struggles in Bolivia unfolded. But it still does not explain the form the aggression took on the Black Day of Sucre; this requires an in-depth analysis of the conflict configuration. I will concentrate on three aspects: (1) street politics, (2) strategic essentialism, and (3), symbolic demonstrations of power.

Street Politics

In Bolivia, collective political action manifests itself mainly as a culture of street politics, bypassing the responsible democratic institutions. This phenomenon applies to several social sectors, in particular to organized labor movements, peasant and indigenous movements, neighborhood organizations, and subaltern sections of the populations, which only form temporary interest groups or convene for specific protests or campaigns. However, the political Right has also chosen the street as their battle field.

Since 2000 street politics (Calderón and Szumkler 2000), including violent protests, have become an ever more inherent part of Bolivian everyday life. However, they have always been part of the nation's political instrumentation. As I have demonstrated in my analysis of Bolivian social movements (Ströbele-Gregor 2008), this form of policy is an important part of the independent political culture of these movements and has a long tradition based on historical experiences (Mayorga 2007: 5). These are based on the insight that previous governments and traditional political parties use public institutions mainly for their own interests, even in times of democracy. Subaltern groups as well as organized labor, miners, peasants, indigenous people, inhabitants of city slums, and others, have seldom been listened to and have often been denied their rights. Self-organization has been, and still is, a survival strategy as well as a prerequisite for political participation. Political action happens in the streets, where it will have an impact and where goals can be achieved.

Non-indigenous citizens as well as the middle and upper classes have always feared and respected the strong assertiveness and mobilization of the socially disadvantaged. Note, for example, the road blocks on arterial roads and access roads erected by local grassroots organizations that paralyzed traffic and hindered supplies in several cities, particularly La Paz. Since the 1990s, the indigenous people of the lowlands have used week-long protest marches from far-away regions into La Paz as an instrument to forcefully demand territory, the acceptance of indigenous rights, and a new constitution. Miners march regularly from the mining centers to La Paz in order to gain more attention for their demands. Setting fire to sticks of dynamite forms part of their symbolic protest. Whether during the "Water War of Cochabamba" in 1999–2000 against the privatization of drinking and irrigation water,[3] the uprising of 2003, or the "water war" in El Alto 2005 against the treaty between the government and an international water company, the political assertion of social movements has happened in the streets, where demands and objections that would not have been listened to in parliament can be heard.

Students have a long tradition in expressing their demands with vocal political protests and militant demonstrations. Violent confrontations with law-enforcement services are the norm. However, supporting demands vocally on the streets is also widespread among the middle classes and party-controlled citizen committees, even though they have direct access to formal policy-making. Ever since the uprising in October 2003, the streets in the departments of the *Media Luna* have been dominated by ultra-right associations and violent youth groups. They consider themselves to be the counter-movement to the demands of the strengthened social grassroots movements of the highlands and the indigenous movements. The exceedingly fascist organization *Unión Juvenil Cruceñista* is active in Santa Cruz. It has close connections to the citizen committee and right-wing parties. These adolescents are predominantly from the middle and upper classes and champion

---

3   In 1999–2000 a spontaneous, broad but temporary alliance formed in Cochabamba including all social classes. In the protest actions, having become famous as the "Water War of Cochabamba" (guerra del agua), citizens that were not usually politically active participated as well as the rural population, unions, and several other city organizations and surrounding communities. The cause was the privatization of drinking and irrigation water in an area that has been suffering from lack of water for years. This evolved into a social movement which concerned itself not only with one particular interest, but with the greater good. The resistance was successful. This was possible because all involved underwent a learning and awareness process in the context of street politics. Spontaneous protest transformed into a strong temporary alliance in which different social groups cooperated in a goal-oriented fashion and managed to stop the government plans (for further reading, see Crabtree 2006).

their claim to a "white" Bolivia, at least in the *Media Luna*. Well-equipped with weapons and the knowledge of their being backed by the political rulers of Santa Cruz, they manifest their racism with violent attacks against the "*indios*."

Considering the background to this political culture of street politics, it is unsurprising that government opponents as well as supporters carry on their arguments about the constitution in the public domain of the streets. Political intolerance on both sides is just as common. Yet, all sides were horrified by the three persons killed during the violent confrontation in November 2007. Horror, shame, and disgust also abounded after the Black Day of Sucre. Traces of these emotions can still be found in newspapers and the internet to this day. Sucre has assumed a new character in the confrontations. "Racist Bolivia," "Ku Klux Klan in Bolivia," "Sucre—Capital of Racism," "Racism and Human Rights Abuse in Bolivia"—these or similar were the headlines of internet entries of Bolivians ashamed for their country: "The occurrences in Sucre are a national shame."

Since the revolution in 1952 and until the tragedy, Bolivia had been known as a country where racism had no place in public discourse. The discrimination and marginalization of people with indigenous names and appearance had been considered by many Bolivians as a social problem which had less to do with racism and more with a lack of education and backwardness in rural areas. Economically successful entrepreneurs among the indigenous migrants in the cities were often used as an example of this theory, as was the vibrancy of the indigenous-*mestizo* cultures. "We are all *mestizos*," "We all have indigenous blood"—these statements were often told to foreigners when they mentioned the latent racism. When in 1993 an Aymara intellectual became vice president, many believed this to be a sign of the interculturality of Bolivian society. Therefore, additional aspects of the political culture in Bolivia need to be examined in order to explain the Black Day of Sucre.

Strategic Essentialism: Strategies of the Powerless in the Political Arena

Since the National Revolution in 1952, the "*indios*," who had been previously socially excluded, have theoretically enjoyed full citizenship. In reality, they continue to be marginalized, discriminated against, and used by political parties as disenfranchised voters. Whereas the miners and labor movement have, since the beginning of the twentieth century, been characterized by pronounced class consciousness. They formed their own organizations, forceful unions, and leftist parties, while the situation in the rural and indigenous hinterlands was quite different. Although life in rural communities has been strongly influenced by indigenous traditions and ancient customs, a coherent political identity did not develop from these commonalities. Even though indigenous representatives nowadays often cite the many regional and local uprisings from before the revolution, these uprisings aimed at the oppression by "white landlords" (*gamonales*). They were not based on the construction of a collective ethno-political identity and an associated political consciousness. By using Gramsci's term of subaltern, we can describe the situation of the rural population as well as the rural migrants in the cities very well up until the 1980s. The political home for the majority of the *campesinos* up until the mid-1980s were the traditional parties, in particular the Revolutionary Nationalist Movement (*Movimiento Nacionalista Revolucionario* [MNR]). The indigenous parties of the 1980s, with their explicit indigenous discourse, were mainly founded by city migrants, mostly Aymara with secondary education, but they were lacking followers in the rural areas.

Nevertheless, since the end of the 1970s, an ethno-political discourse developed within the organized peasant movement *Movimiento Katarista* in the Andean highlands. The indigenous people of the lowlands followed suit a couple of years later. The indigenous parties did not develop any appreciable popularity in either the highlands or the lowlands. For the rural population, cultural identity amounted to local identity. There was no comprehensive "We-group-consciousness" and even less a common sociopolitical project (Ströbele-Gregor 1994a). The ethnopolitical discourse remained a discourse of self-proclaimed "indigenous representatives." It did not have a great deal in common with the self-image of the population. The ruling populist ideology of "*el pueblo*," *mestizaje*, and social integration as heritage of the National Revolution was too influential until 1985, despite the fact that this ideology had nothing to do with reality. The ideological alternative was characterized by class discourse. Communicated through the labor movement and leftist parties, but with regional differences, it was absorbed in some parts of the *campesino* community who identified themselves as *trabajadores campesinos* (peasant workers). The class discourse blended with discourses of cultural identity. Outstanding examples of this phenomenon include the umbrella organization of the *campesino* unions *Confederación Sindical Única de Trabajadores Campesinos de Bolivia* (CSUTCB), the movement of the Coca farmers, and—several years later—Evo Morales and his movement party MAS.

In 1985, the situation culminated in a final break with the ideology of the Revolutionary National State of 1952 and its promise of integrating the excluded and marginalized people. The government of the former Revolutionary Party MNR started to modernize society within the framework of a strictly neoliberal policy. Miners, workers, and the rural population suffered the brunt of the policy shift. The neoliberal withdrawal of the state unwillingly promoted the political organization of the excluded, in particular the indigenous rural population (Ströbele-Gregor 1994b). The change in the relationship, especially of the rural population, to the state coincided with an altered self-image of the social protagonists in the political arena and with the capacity to act. They were looking for connections and common ideals to emphatically represent their own interests and to find common positions against those whom they considered responsible for social inequality and political exclusion: the economic power groups and their political instruments, the traditional parties. It were predominantly the descendants of the small European-rooted middle and upper classes who had always held political power in their grasp. Their counterpart was the majority of the "non-white" population. Politics were more and more often expressed in these terms. The term "strategic essentialism," coined by Gayatri Chakravorty Spivak, is especially suited to the analysis of the political actions of the social and indigenous movements in Bolivia at the beginning of the 1990s. In her analysis of an anti-essentialist intended historiography Spivak makes out latent essentialist attitudes in her reconstruction of subaltern experiences in colonial India. She ascribes them a strategic and emancipatory function. Following that thought will allow one to create a strategy for assertion in the political field (1996).

The, to some extent, culturally very different ethnic groups started to slowly articulate the difference between them and the people of European descent by constructing essentialist discourses, which in turn enabled them to find similarities and a new self-esteem. Despite the great cultural diversity in Bolivia, which for a long time had also entailed mutual differentiation, it was now possible to construct a new political identity as "indigenous people" via an "indigenous culture." It allows for the organization of political action on a regional scale and also serves to justify the groups' demands to the state. The call for a constitutional assembly with equal participation of indigenous people has become a lynchpin of this strategy. The fundamental criticism of the prevailing, extremely unjust conditions, oppression, and the destruction of livelihoods gained a new political quality. The demands of individual groups were no longer the main concern. With the construction of a collective "we," the formerly excluded "indigenous people" legitimized their demands for fundamental social and political change. When these demands gained more support and Evo Morales, a representative of this position, was elected as president, the former power and ruling groups must have felt considerably challenged. For the first time since the National Revolution of 1952, their privileges and predominance were in serious danger.

Symbolic Demonstration of Power

The more openly racist rhetoric of the power groups as well as their associated social sectors and political representatives should be seen in this context. The uprising of October 2003 in El Alto and La Paz was a signal to the former leaders. The uprising, as has already been mentioned, was carried out by the indigenous-mestizo ethnic groups. It ended with the resignation and exile of President Gonzalo Sánchez de Lozada. In President Sánchez de Lozada, a prosperous mine-owner who had further been raised in the United States, the social and indigenous movements saw an ideologist and protagonist of the neoliberal restructuring of the state, a symbolic figure of the "white power groups." His forceful banishment was a more fundamental challenge to these power groups than the dispute of current economic political programs. The ruling classes felt threatened in their self-image and their social position as "whites," as "elites" with a hereditary right to rule. They reacted with total opposition against the government on all levels, on the one hand, and by ideologically legitimizing their claim to rule and autonomy, on the other. They constructed a self-image as *criollos,* the "purebred" heirs of the Spanish *conquistadores,* who had conquered the lowland regions, or as descendants of European migrants, and thus justified their claims to superiority. Their demands of a self-governed territory and rejection of the central government (Plata 2008) can bee seen as a result of this way of constructing history. The native indigenous people only feature in this image as cheap labor. Correspondingly, they ascribe economic successes in this region to the *criollos'* drive and genes. Racist slogans against the "*indios*" of the poor Andean departments have their ideological roots here. In this view, giving power to "*indios*" or allowing them to participate politically means to relinquish the country to the less civilized, the uneducated, the inept, and the inferior. The violent racist attacks on "*indios*" or the torching of the offices of indigenous organizations should be seen as political signals, as a fight against all identified as "*indios*."[4]

The violent acts of abasement in Sucre recall the punitive acts of the colonial period and also the rule of the landlords before the revolution of 1952 (Defensor 2009: 32). These acts were staged as a symbolic reminder that the old hierarchical social order and the position of power of the former rulers still exist. The timing could not have been clearer: the celebrations of the first uprising against colonial rule, *el primer grito de Libertad en Sucre,* an event which unmistakably emphasizes the historic claim of dominance the *criollos* demand of the state borne out of the independence wars. The violent perpetrators use the high symbolic value of the time and place to convey their message. The presence of the political authorities of Chuquisaca as well as the media, who were gathered for the commemorative celebrations, was an important component of these staged acts. In this way, the protagonists could be sure that their symbolic re-establishment of the old order and position of power would be "witnessed" by the whole country, serve as a warning to the "*indios*," and bolster the self-confidence of the "rulers."

The protagonists were mostly adolescents and students. The incident raises the question of how those individuals who were born such a long time after the revolution could develop such notions of colonial forms of demonstrating power and social order. An internet entry mentions the socialization of the middle and upper classes: from an early age onwards, these children know the "*indios*" as housemaids, servants, and gardeners. They see the dark-skinned poor begging on the streets. When they travel into the countryside, they see the *campesinos* in their miserable houses, whereas they only meet children like themselves in their private schools. All these experiences lead to a feeling of superiority deeply ingrained in the consciousness of these social classes. Inequality, a feeling of superiority, and open or hidden racist convictions or at least contempt have deep roots in the mentality of the middle and upper classes, especially in the *Media Luna.* Of course, as internet entries and many personal statements prove, not everyone shares these attitudes.

---

4   The fact that the government seemed unwilling to use force can be explained, in my opinion, by the unwillingness of Morales to allow the violence to escalate. Massive police forces would in all likelihood strengthen the willingness of local anti-government groups to use force, but this aspect cannot be looked into here.

However, a 2008 investigation by the human rights organization UNIR concerning the press coverage of the Black Day clearly reveals that the media also need to be more conscious of the actual meaning of being a citizen. In the eight newspapers investigated, the ideological reasons, mindsets, and motives of the protagonists were hardly covered at all, and citizenship and what this entitles seemed somewhat unclear. The newspapers noticed antagonist "enemy groups" and framed their confrontation as that of "*campesinos* against citizens" or "adolescents against *campesinos*," leading UNIR to the conclusion that, apparently, these newspapers do not consider the *campesinos* citizens (UNIR 2008: 11).

Conclusion

The UNIR's findings are consistent with the self-perception of the *criollos* and their construed "white city," in which the "*indios*" are *campesinos*, peasants. As such, their living space is viewed as being located outside of the city; indigenous people are thus considered neither as citizens of the city nor as belonging in the city (Defensor 2009: 27). At best, they are tolerated as folkloristic tourist attractions. With this mental cartography, the colonial cultural heritage of the "white city" remains vivid, and the colonial image of society which constructs a clear ethnic separation and segregation is reinforced. The *indios-mestizos* are socially defined as "under classes" in the urban space: as service workers, traders, and the poor. It appears as though they should remain as far as possible in the shadows of the everyday image of the city for fear that their visibility would ruin its beauty. Furthermore, persons with an indigenous background who live in the city and who, qua education or economic success, have risen to a certain degree in the social hierarchy, remain excluded from the "high society" of the *criollos*, even when they attempt to fit in through strategies of "whitening."

Nonetheless, this ethno-social hierarchy, including the exclusion of the "*indios*" from and within the city, is based not only on the continuing existence of a symbolic order rooted in the colonial period; it is a habitus which was also heavily influenced by the so-called scientific racist ideas of the second half of the nineteenth century. Demelas (1981) has impressively discussed the influence of these ideas on the interpretation patterns of the dominant social groups and their intellectuals in Bolivia up until the beginning of the twentieth century. Above and beyond this, however, the mixture of a colonial symbolic order and of *racismo a la criolla* is yet to be overcome in the mindset that shapes the image of society of the *criollos*, as the report by the United Nations' special rapporteur Stavenhagen (2007; also "El rebrote" 2007) ascertains. According to tenor of the study by the *Defensor del Pueblo* (Ombudsman for Human Rights) (2009), the "Black Day" of Sucre offers proof that this is the case. Building a multinational state which is based on respecting and acknowledging the diversity and equal rights of all citizens in Bolivia appears to be more difficult than imagined. The colonial mentality and the will to rule seem to be far too widespread within the middle and upper classes.

Works Cited

Albó, X. 2002. *Pueblos indígenas en la política.* La Paz: Plural.

Albó, X. 2008. La iglesia y los esclavos guaraníes. *infoderechos.org* [Online]. Available at: http://infoderechos.org/es/node/129 [accessed: April 1, 2009].

Bedoya, E., and Bedoya, A. 2005. *El régimen de servidumbre y enganche por deuda.* [Online]. Available at: http://se1.isn.ch/serviceengine/FileContent?ser viceID=47&fileid=A34FCC4A-7024-1679-026E-BC2ABA962A00&lng=es [accessed: April 1, 2009].

Calderón, F., and Szmukler, A. 2000. *La política en las calles.* La Paz: CERES/ Plural.

Chávez, F. 2008. Humillación indígena obliga a cambiar. *Agencia de Noticias Inter Press Service* [Online, May 28]. Available at: www.ipsnoticias.net/nota. asp?idnews=88566 [accessed February 22, 2009].

ComunicaBolivia. 2008. *Bolivia: 18 campesinos son vejados y humillados en Sucre.* [Online, May 26]. Available at: http://www.youtube.com/ watch?v=5RXUkPrYHcE [accessed: February 22, 2009].

Contreras, A. 2008. Facismo racista en Bolivia. *Chamosaurio. Actualidad política de Venezuela.* [Online, May 26]. Available at: http://chamosaurio. com/2008/05/26/fascismo-racista-en-bolivia/#more-2949 [accessed: March 20, 2009].

Crabtree, J. 2006. Patterns of Protest: The Genesis of a New politics in Bolivia, in *Bolivien – Neue Wege und alte Gegensätze,* edited by F. Bopp and G. Ismar. Berlin: Wissenschaftlicher Verlag, 151–82.

Defensor del Pueblo Bolivia, and Universidad de la Cordillera. 2009. *Observando el racismo. Racismos y regionalismos en el proceso autonómico. Hacia una perspectiva de clase.* La Paz: Canasta de Fondos.

Demelas, M. 1981. Darwinismo a la criolla: el darwinismo social en Bolivia, 1880–1910. *Historia Boliviana,* 1, 55–82.

Deutscher Entwicklungsdienst and Equipo Técnico del Programa "Fomento al Diálogo Intercultural en el Chaco Boliviano." 2008. *Familias guaraníes empatronadas: Análisis de la conflictividad.* La Paz: Servicio Alemán de Cooperación Social Técnica.

Goedeking, U. 2002. Die Macht politischer Diskurse: indigene Bewegung, lokaler Protest und die Politik indigener Führungspersönlichkeiten in Bolivien, in *Dossier: Nuevas tendencias de los movimientos indígenas en los Países Andinos y Guatemala a comienzos del nuevo milenio,* coordinated by J. Ströbele-Gregor. Berlin: INDIANA Bd.17/18, 83–104.

Guha, R. 1982. Preface, in *Subaltern Studies I: Writings on South Asian History and Society,* edited by R. Guha. Delhi: Oxford University Press.

Instituto Nacional de Estadística de Bolivia. 2001. Autoidentificación con pueblos originarios o indígenas de la población de 15 años o más de edad—ubicación, área geográfica, sexo y edad, en *Censo 2001 [Población].* [Online]. Available at: http://www.ine.gov.bo:8082/censo/make_table.jsp?query=poblacion_06 [accessed: July 8, 2010].

Klein, H.1984. *Historia General de Bolivia*. La Paz: Ed. Juventud.

Mayorga, F. 2007. Movimientos sociales, política y estado. *ConstituyenteSoberana. org* [Online]. Available at: http://constituyentesoberana. org/3/ docsanal/072009/280709_1.pdf [accessed August 20, 2009].

Oficina del Alto Comisionado de las Naciones Unidas para los Derechos Humanos–Bolivia. 2009. *Informe Público de la oficina del Alto Comisionado de las Naciones Unidas para los Derechos Humanos de Bolivia sobre los Hechos de la Violencia Ocurridos en Pando en Septiembre 2008*. La Paz: Naciones Unidas.

Piepenstock, A., Vargas, G., and Goedeking, U. 2004. Vom Musterland zum Volksaufstand: die Krise des neoliberalen Modells in Bolivien, *Jahrbuch Lateinamerika: Analysen und Berichte*, 28, 142–53.

Plata, W. 2008. El discurso autonomista de las élites de Santa Cruz, in *Los barones del Oriente: El poder de Santa Cruz ayer y hoy*, edited by X. Soruco, W. Plata and G. Medeiros. Santa Cruz: Fundación Tierra, 101–72.

Platt, T., Martínez, R., Rivière, G. 2008. Racismo y violación de los Derechos Humanos en Sucre, Bolivia. *Banco tematico* [Online]. Available at: http://www. bancotematico.org/archivos/primeraMano/archivos/racismo_en_bolivia.pdf [accessed: February 25, 2009].

Presidencia de la República de Bolivia. 19 de Noviembre 2008. *Decreto Supremo Nro. 29802*.

Postero, N. 2007. *Now We Are Citizens: Indigenous Politics in Postmulticultural Bolivia*. Stanford: Stanford University Press.

Spivak, G. C. 1996 [1985]. Subaltern studies: deconstructing historiography, in *The Spivak Reader*, edited by D. Landry and G. MacLean. London: Routledge, 203–36.

Ströbele-Gregor, J. 1994a. From indio to mestizo ... to indio: new indianistic movements in Bolivia. *Latin American Perspectives*, 21(2), 106–23.

Ströbele-Gregor, J. 1994b. Abschied von Stief-Vater-Staat: wie der neoliberale Rückzug des Staates die politische Organisierung der Ausgeschlossenen fördern kann. *Jahrbuch Lateinamerika: Analysen und Berichte*, 18, 106–30.

Ströbele-Gregor, J. 2006. Für ein anderes Bolivien—aber für welches? Indigene Völker und Staat in Bolivien, in *Bolivien- Neue Wege und alte Gegensätze*, edited by F. Bopp and G. Ismar. Berlin: Wissenschaftlicher Verlag, 279–326.

Ströbele-Gregor, J. 2007. Bolivien im Umbruch: ein Jahr Evo Morales—eine Zwischenbilanz *Jahrbuch Lateinamerika: Analysen und Berichte*, 31, 182–93.

Ströbele-Gregor, J. 2008a: Kanon mit Gegenstimmen: soziale Bewegungen und Politik in Bolivien, in *Jenseits von Subcomandante Marcos und Hugo Chávez: Soziale Bewegungen zwischen Autonomie und Staat: Festschrift für Dieter Boris*, edited by S. Schmalz and A. Tittor. Hamburg: VSA, 129–41.

Ströbele-Gregor, J. 2008b. Bolivien 2008: Spiel mit dem Feuer: Bericht September 2008. *Heinrich Böll Stiftung* [Online]. Available at: http://www.boell.de/weltweit/lateinamerika/lateinamerika-4852.html [accessed: February 22, 2009].

Ströbele-Gregor, J. 2009. Kampf um Land, in *Die Neugründung Boliviens? Die Regierung Evo Morales*, edited by T. Ernst and S. Schmalz. Baden- Baden: Nomos, 141–54.

UNASUR. 2008. *Informe de la Comisión de UNASUR sobre los sucesos de Pando. Noviembre 2008.* [Online]. Available at: www.cedib.org/bp/Infunasur. pdf [accessed: February 22, 2009].

UNIR. 2008. Violencia en Sucre: ¿quiénes fueron los protagonistas de las noticias? *Puertas Abiertas,* 4(2): 11.

PART II
The Use of Ethnicity in the Imagineering of Urban Landscapes

Introduction to Part II

The Use of Ethnicity in the Imagineering of Urban Landscapes

Olaf Kaltmeier

Paralleling cultural studies and sociosemiotic approaches, in which the city is understood as text and image (Gottdiener 1998, Gottdiener and Lagopoulos 1986, Duncan 1990), discussions of the city in post-Fordist urban cultural politics and city marketing focus on terms like "image city," "theme city," and the "city as stage" for cultural events. Here, the relationship between city, image, and identity is of particular importance. In this sense, Denis Cosgrove understands landscape as technically and epistemologically produced "ways of seeing" (1998: xv), whereas John Urry, following Michel Foucault's notion of the "clinical gaze," conceives of the touristic perception of space as the "tourist gaze" (1990). Even though Martin Jay has already described modernity as an "ocularcentric" (1993) age, one can witness the further acceleration of the circulation of images during the information age (Castells 1996, 1997), consequently leading to an image of the city which is increasingly conveyed via the media (Martín Barbero 1987).

Although the planned experience of city and landscape has its historic predecessors—for example in the landscape gardens of the nineteenth century—this dimension of experiencing city and countryside has gained importance in post-Fordist consumer societies. "Imagineering," a neologism the Disney Corporation created by blending the words "imagination" and "engineering," most clearly articulates the technically constructed and planned consumability of spaces and identities. In this concept, architecture functions in particular as a façade, as a stage for cultural spectacles. The reference to the Disney Corporation here is not merely metaphorical, as the company, in rebuilding New York's Times Square as well as in establishing the gated community "Celebration" in Florida, has been pursuing urban development programs of its own which are tailored to target precisely defined lifestyle groups.

One of the key differences between these kinds of projects and their predecessors lies in the fact that imagineering aims economically to stimulate consumption and hence to make profits. For the last third of the twentieth century, a proliferation of themed and imagineered urban ensembles can be witnessed, from theme parks via shopping centers and gated communities to the re-semantization and revitalization of historically grown urban districts, especially "ethnic" neighborhoods, parks, central squares, or historic city centers. Fundamental changes in the use of the architectural substance accompany the aforementioned processes, as the branches of transnational corporations as well as the offices of the creative class increasingly drive downtown lodgings, tradesmen's workshops, and retail stores out of their traditional locations. These developments influence the image of cities in that people's mental maps of a given city become increasingly removed from their material and corporally experienced reality and, thus, the city consequentially becomes a simulation of itself.

On the one hand, one may conceive of these processes in the sense of a convergence and emergence of a "mimetic isomorphism" (DiMaggio and Powell 1983). They can be understood as symptoms of a tendency to standardize urban cultures that manifests itself especially in the dissemination of "non-places" (Augé 1994) such as shopping centers or airport terminals as well as in the convergence of cultural and industrial places, for example in increasing Disneyfication processes. On the other hand, one may identify a progressive multiculturalization and hybridization of cities that has led to the formation of diverse so-called "urban ethnic spaces" within cities that seek to distinguish themselves from other cities and create unique selling points of their own. Diversity marketing is becoming a key term of the new urban cultural politics. Following Slavoj Žižek (1997), one can criticize the recent hype of cultural and ethnic diversity as only an *Erscheinungsform* (manifestation of a reality) based on a culturally homogenizing and globally spreading capitalism.

Reaching beyond the production of the city as image and its media circulation, the question of its reception arises. In his path-breaking study *The Image of the City* (1960), Kevin Lynch worked out universal structural elements for the perception of cities. Conversely, the Latin American urban researcher and co-founder of Latin American *Estudios culturales*, Néstor García Canclini, analyzes different patterns of perception in the city, which he conceives of as "urban imaginaries." He writes:

> We not only experience the city physically, we not only walk around and feel in our bodies what it means to walk for a certain time or stand in a bus or to be out in the rain until one manages to find a taxi, but we also imagine while we travel; we make assumptions about what we see, about those who cross our way. ... A great part of what happens to us is imaginary, because it does not emerge from a real interaction. Every interaction has an imaginary portion, and this applies the more to those evasive and fleeting interactions a megacity brings about. (García Canclini 1997:88–89, own translation)

Especially in the megacities, urban space can no longer be adequately understood as a functional organic unit. Instead, one can witness how cities fall apart into different subsystems that even the inhabitants no longer perceive as an integrated whole, as, for example, studies on Mexico City (García Canclini 1995, 1997) have shown. One may speak of a "video clip city" (García Canclini 1997: 88) rather than of a shared, integrated image of the city.

However, it is this fragmentation of the single, integrated image of the city, in particular, that creates space for ethnic diversity. Fordism defined the image of the city principally in terms of a motor of industrial progress and as a symbol of the nation, including the processes of de-ethnization and cultural assimilation tied to these notions. Understanding the city as a video clip on the one hand implies a multiplication of regimes of the gaze that may render ethnic diversity particularly visible. On the other hand, the metaphor of the video clip points to the fleeting nature of representations tailored to enable immediate consumability.

In "Urban Landscapes of Mall-ticulturality," Olaf Kaltmeier relates the construction of the San Luis Shopping Center in Quito, a mall designed in a neocolonial *Hacienda* Style, and its transformation of the urban landscape to the broader social context of identity politics in Latin America. His main theses are that the San Luis Shopping Center exemplifies a wider process of "retro-colonizing" the urban landscape, that the construction of the mall articulates the recent tendency in urban living to return to the city center, and that San Luis forms part of an identitarian-spatial segregation that takes the form of a fractal archipelago.

Julie TelRav discusses the reciprocal relationship between architectural environments and the development of a religious and cultural community. By looking at the architecture and environmental design of two buildings of the Jewish Community Center (JCC) in the Detroit Metropolitan area, her chapter works out the underlying strategies which the Detroit Jewish community employs in order to maintain its historic ties to religious and cultural tradition, on the one hand, and to adapt to the changing needs of the community today and in the future, on the other. The JCC makes for an especially fruitful, and, in some ways, comparative case study, as the two buildings studied here, while serving similar basic functions, are located in two highly distinct neighborhoods and hence cater to widely divergent constituencies.

Ruxandra Rădulescu analyzes the re-imaginings of urban-indigenous identities in twenty-first-century Seattle. Signaling his trickster discourse of in-betweenness, the phrase "ambiguously ethnic," used by Spokane writer Sherman Alexie in *Ten Little Indians* (2004), is a controversial (self-deconstructing) label that challenges received or desired notions of ethnic authenticity, particularly within an urban geography. Rădulescu's chapter argues that the multiple cultural allegiances of Alexie's characters are not a sign of naïve cosmopolitan idealism, but instead follow a logic of contestation and "violence of hybridity." In so doing, the narrative landscape / urban ethnoscape that they construct is fraught with challenges, dramatic failures, and parodical play with popular culture stereotypes in a "postindian" redefinition of community.

Jens M. Gurr and Martin Butler are particularly concerned with the medial circulation of urban imaginaries. Using Norman Klein's multimedia documentary *Bleeding Through: Layers of Los Angeles 1920–1986* as a case study, they explore the role which hypertext docufiction can play in the cultural representations of urban complexity and multiplicity. Taking theirs cue from Cultural Memory Studies, Ecocriticism, and the study of social sustainability, they argue that the genre of multimedia docufiction is uniquely suited as a medium of memory and of endowing the hitherto often unheard voices of ethnic minorities in the city with a voice of their own. In so doing, their chapter also suggests ideas pointing toward a model of classifying forms and functions of urban cultural practices according to the mediacy of their involvement with urban space.

Works Cited

Augé, M. 1995. *Non-Places: Introduction to an Anthropology of Supermodernity.* London: Blackwell.

Castells, M. 1996. *The Information Age: Economy, Society and Culture*, vol. 1: *The Rise of the Network Society.* Oxford and Malden, MA: Blackwell.

Castells, M. 1997. *The Information Age: Economy, Society and Culture*, vol. 2: *The Power of Identity.* Oxford and Malden, MA: Blackwell.

DiMaggio, P.J. and Powell, W. 1983. The iron cage revisited: institutional isomorphism and collective rationality in organizational fields. *American Sociological Review*, 48(1), 147–60.

Duncan, J. 1990. *The City as Text: The Politics of Landscape Interpretation in the Kandyan Kingdom.* Cambridge: Cambridge University Press.

García Canclini, N. 1995. Mexico: cultural globalization in a disintegrating city. *American Ethnologist*, 22(4), 743–55.

García Canclini, N. 1997. *Imaginarios urbanos.* Buenos Aires: Editorial Universitaria de Buenos Aires.

Gottdiener, M. 1998. *Postmodern Semiotics: Material Culture and the Forms of Postmodern Life.* Oxford: Blackwell.

Gottdiener, M. and Lagopoulos, A. (eds.) 1986. *The City and the Sign: An Introduction to Urban Semiotics.* New York: Columbia University Press.

Jay, M. 1993. *Downcast Eyes: The Denigration of Vision in Twentieth-Century French Thought.* Berkeley: California University Press.

Lynch, K. 1960. *The Image of the City.* Cambridge, MA: MIT Press.

Martín Barbero, J. 1987. *De los medios a las mediaciones: Comunicación, cultura y hegemonía.* Naucalpan: Gili.

Urry, J. 1990. *The Tourist Gaze: Leisure and Travel in Contemporary Societies.* London: Sage.

Žižek, S. 1997. Multiculturalism, or, the cultural logic of multinational capitalism. *New Left Review*, 225, 29–49.

Chapter 5

Urban Landscapes of Mall-ticulturality: (Retro-)Coloniality, Consumption, and Identity Politics: The Case of the San Luis Shopping Center in Quito

Olaf Kaltmeier

When driving down the highway from the center of the Ecuadorian capital, San Francisco de Quito, towards the Chillos Valley, which is one of the main middle-class suburban residential zones, one can see upon first glance the white, glaring, impressive building of a new mall, the San Luis Shopping Center. The construction of this mall was concluded in September 2006. At 43,000 square meters, with 150 shops and ten cinemas, it is one of the biggest shopping centers in Latin America. Nevertheless, the most outstanding characteristic is not its size but its architecture, which is linked to aspects of cultural heritage, "authenticity," and identity politics. The special aesthetics of the San Luis Shopping Center may be best expressed by a statement of the jury of the 32nd Design and International Development Award 2008 of the International Council of Shopping Centers (ICSC):[1]

> Historically, the most important haciendas of Ecuador were located in Los Chillos Valley. Nowadays, this tradition has been recovered through the development of an amazing mall called San Luis Shopping, which is known as the commercial heart of the valley. Architecturally, the mall maintains the façades of the Andean haciendas with wide walls, heavy ceilings, hand-forged iron, and thousands of details that belong to the Spanish Colonial period. ... It will be the first project in South America that merges history, business, and lifestyle. Two hundred years ago, Los Chillos Valley was the heart of a blooming area of beautiful haciendas. ... Nowadays, this part of our history is being preserved through the

---

1 Founded in 1957, the ICSC is the global trade association of the shopping center industry. Its 70,000 members in the United States, Canada, and more than eighty other countries include shopping center owners, developers, managers, marketing specialists, investors, lenders, retailers, and other professionals as well as academics and public officials. As the global industry trade association, ICSC links with more than twenty-five national and regional shopping center councils throughout the world.

construction of an amazing mall that is considered, by the community, as the heart of the valley: San Luis Shopping. (2000)[2]

This nostalgic use of colonial and local elements in order to produce authenticity was honored internationally by the gold medal in the category of innovation in design and development of the International Council of Shopping Centers in 2009.

This use of culture to stimulate urban development and consumption is characteristic of the post-Fordist cultural politics (Zukin 1995, García Canclini 1995) to an extent such that some critics such as George Yúdice (2003) have even argued that an expediency of culture has become epistemological in this period. What is striking in the case of the San Luis Shopping Center is the fact that a concept as controversial as coloniality—and in particular the hacienda—is used as a mode of marketing and branding at the beginning of the twenty-first century.

Only few segments of the Ecuadorian population would respond in an explicitly positive manner to this concept of colonialism, which highlights Hispanism as a real and imagined shared heritage with European culture, which is conceptualized as being superior to the indigenous cultures. In general, coloniality is related to connotations of exploitation, conquest, and domination and thus to feelings of colonial shame. Such associations are particularly characteristic of the Andean hacienda, which in the nineteenth century became the main locus for the control and exploitation of the indigenous populations (Kaltmeier 2011). In Ecuador in particular, the hacienda was such an important dispositive that in the second half of the nineteenth century the country could be understood in terms of a "hacienda society" (Quintero 1986). In the mid-twentieth century, the social imaginary changed, and the hacienda became a symbol of "backwardness" and "barbaric traditions" that had to be overcome. Within the literary *indigenismo*, Jorge Icaza's influential work *Huasipungo* best described the brutal oppression under the hacienda regime. Finally, the agrarian reforms in the 1960s and 1970s put an end to the hacienda and its system of debt bondage or peonage

The statement by the ICSC emphasizes the idea that the mall merges "history, business, and lifestyle" (2008). Thus it becomes clear that the success of the hacienda mall lies precisely in the appellation of certain identities and places. It can be argued that colonial aesthetics are an integral element of identity politics related to the urban middle- and upper-class sectors around Quito which the San Luis Shopping Center explicitly targets (*El Comercio*, October 12, 2006; Interview with Carmen Recalde, January 31, 2008). With the failure of the nation-building project based on assimilation, *blanqueamiento* (striving toward whiteness), and white supremacy as well as with the end of the hegemonic discourse of

---

2   The first hacienda-style shopping center in Latin America was erected in Bogotá in the context of the renovation of the Santa Barbara Hacienda in 1987, which was declared a national heritage in the same year. A conversion of the hacienda into a mall was planned and realized in 1990. (Centro Comercial Hacienda Santa Barbara n.d.).

modernization and development predominant throughout most of the twentieth century, the urban upper and middle classes have lost their position of leadership in the definition of the principles of vision and division of the social world. Since the 1990s, indigenous sectors have instead achieved a pluricultural redefinition of the nation through a process of the "ethnicization of the political" (Büschges and Pfaff-Czarnecka 2007, Kaltmeier 2007). The economic crisis (for example the dollarization of the national economy) at the end of the twentieth century further reinforced the ideological identity crisis of the white middle and upper classes.

In this moment of deep identity crisis, a nostalgic recourse to coloniality as a secure and stable element seems to offer a promising solution. However, this course is not to be understood as a simple return; instead one can state a new articulation of identity for which consumption and lifestyle are as important as notions of authenticity and *Heimat*. After an introduction in sociosemiotics and qualitative methods of social analysis in urban landscapes, this chapter will explore the strategic use of coloniality and rurality as well as their commodification, introducing the concept of retro-coloniality. Finally it will discuss the transformations of the urban geopolitics of identity, locating the San Luis Shopping Center in a broader panorama of the dialectics between space and identity politics.

Reading Urban Landscapes: Sociosemiotics and Qualitative Methods

The construction of urban entertainment centers in the Americas is usually closely related to processes of "imagineering," originating from the "dream factory" of the Walt Disney studios. Here, theming is a central aspect, as it relates distinct elements in order to narrate homogeneous stories for a clearly identified public. This practice reveals the overwhelming importance of visual signs in Western consumer societies (Aitken and Craine 2005), where consumption has not only a materialist function in terms of the satisfaction of needs, but a symbolic dimension (Baudrillard 1981, Debord 1995) that is linked to status and distinction (Bourdieu 1984). Hence, a sociosemiotic approach that takes into account the visual may be useful to the analysis of the shopping center and its effects on identity politics and the urban landscape.

Semiotics entails the study of signs and sign systems or texts, and it is an integral part of Cultural Studies, where it is used to analyze regimes of representation (Hall 1997). These sign systems may refer to written words but also to other cultural formations such as (urban) landscapes (Cosgrove 1998, Duncan 1990) or architecture (see the contributions in Gottdiener and Lagopoulos 1986). Semiotics focuses primarily on analyzing how certain objects or elements are taken as signs (signifiers) that express a culturally defined meaning (signified or content) in the process of signification. While semiotics is often merely interested in the denotation, the relation between signifier and signified, for the purpose of this chapter it seems fruitful to include the dynamics of connotation: the sociosemiotic interpretation of the ideologies, imaginaries, values, and myths that specific signs

evoke. Mark Gottdiener has presented an elaborated sociosemiotic model in order
to articulate texts, ideologies, and material culture (1995), which he also applied
to the analysis of shopping centers and theme parks.

Despite their usefulness, orthodox semiotic models also have their limits. First,
there is a strong tendency to conceive of texts as fixed structures, neglecting the
performative process of signification, the signifying practices in the terms of Stuart
Hall (1997). According to Hall, texts are produced by social actors rather than being
a given entity. Taking this into account, one must further keep in mind that urban
landscapes and buildings need to be treated as very special form of texts. A high
amount of economic and cultural capital—in terms of the disposal of technical
knowledge and political capital—is required in order to build and hence to "write"
these texts. Due to this fact, most iconic buildings express ideas and values of the
hegemonic social sectors, and they must be understood as the materialization of
identities in spaces, which is a central aspect of symbolic power.

Secondly, many semiotic approaches neglect the work of reception or decoding.
However, as all texts are polysemic, there is never only a single dimension to read
a text. The construction of meaning instead depends on the decoders' position in
the field of identity politics, mainly with regard to their social class, ethnicity,
gender, age, and consumer character (Gottdiener 1986). Stuart Hall's reception
theory distinguishes among dominant (or preferred), oppositional, and negotiated
readings. In a dominant or preferred reading, the audience perceives the text in the
very manner intended by the creator. In an oppositional reading the audience rejects
the preferred reading and instead creates its own meaning of the text. Negotiated
reading refers to a compromise between the dominant and oppositional readings,
in which the audience partially accepts and partially rejects the author's views.
With regard to urban landscapes, one can apply these concepts to ways of seeing
as well as to spatial practices and affective experiences.

The third limitation concerns the epistemological viewpoint of the researcher,
which is a blind spot in semiotics. One often reads interpretations that position their
authors in an almost god-like position outside the field of signification. The implicit
assumption is that Western academics possess the analytical tools necessary to
deconstruct texts and their meanings in a universalistic way. However, postcolonial
thinkers such as Walter Mignolo have shown the importance of a "geopolitics of
knowledge" (2000), the production of knowledge in which depends on the locus of
enunciation. Academic researchers, despite their more elaborate analytical tools,
are only one type of interpreter of a text among others.

These critical remarks also imply certain methodological reorientations of the
traditional modes of semiotic interpretation—shifts I will discuss in the following
description of my research design.[3] The first part of the research consists of a
quantitatively-orientated visual content analysis (Hopkins 1998: 69), for which it

---

3   The research was carried out in the context of a seminar that I held with students
at the Universidad Andina Simón Bolívar in Quito in 2008. I thank Cristina Ahassi, Pablo
Ayala, Andrés Castro, Mónica Delgado, Ana María Egas, Pablo Larreategui, David Lasso,

was important to detect the relevant architectural and aesthetic features of the mall, designed by the company Ekron Construcciones and its architect Jaime Viteri. This was carried out in the form of guided observations and a data entry form. In a second step the visual perception and documentation in the form of digital photos was codified and, in a third step, categorized in working sessions. Afterward, smaller working groups interpreted all categories and their corresponding codes.

A qualitative data collection completed this "positivist" data collection of the relevant signs of the production of the mall's meaning. Taking into account that multiple interpretations (decodings) of a text (in this case the San Luis Shopping) exist, the research team conducted semi-structured interviews with consumers in the mall, shop owners, neighbors, developers, and municipal authorities in order to explore the different meanings that they attribute to the new Shopping Center.

In order to reduce the problematic aspect of the locus of enunciation and the problem of hermeneutics, we conducted the investigation in a working group characterized by cultural, gender, and regional differences. The mall was not only codified according to the reading of the documentary photographs, but anthropological field research as well as participant observation in the mall formed integral parts of the research process and were recorded in a research chart. In order to work out multiple interpretations, we presented individual interpretations which were then discussed in the research team in an attempt to work out additional possible connotations. The intention of the following section of this chapter is not to present the results in detail; instead my aim is to work out the impacts of the mall on urban development and for the marketing of place and identity. Despite these methodological considerations, the San Luis Shopping Center and its insertion in the field of identity politics in Ecuador are presented here from the perspective of the author of this contribution.

Retro-Coloniality and Landscaping

The San Luis Shopping Center alters the structure and perception of the landscape significantly. At first glance, it is the façade of the mall and its exterior expression that produce its dominating impact on the valley's landscape. The main effect can be described as visibility through bigness (Koolhaas 1995): the mall can be seen from a long distance when coming down the highway from Quito to the Chillos Valley. A huge area of 60,000 square meters is enclosed by a white wall with lanterns, a typical element of both the Andean hacienda and the "architecture of fear" (Ellin 1997) of citadels such as the emerging gated communities.

Upon first inspection, the use of colonial architectural elements and ornaments is striking. However, the exterior design of the mall varies in accordance with its two key functions. Viewed from down the valley from Quito, the main façade

Ana María Pozo, James Rodriguez, Alex Schlenker, Elisabeth Solano, Verónica Terán, and Geovanny Villegas for their contributions.

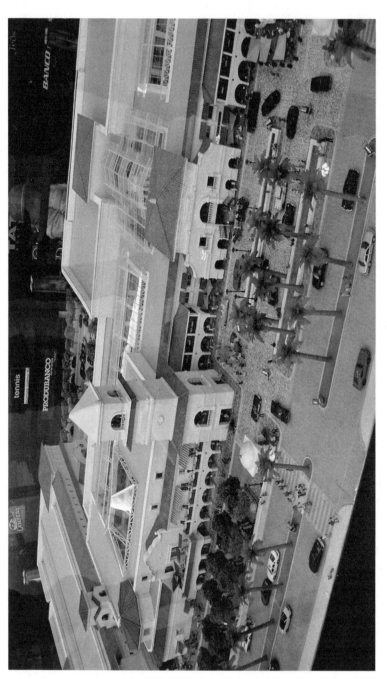

Figure 5.1   Model of the San Luis Shopping Center, Chillos Valley, Ecuador. Seen from the back. Photograph by the author

and entrance of the building focus on the functional aspect of selling. Only a few windows and balconies interrupt the brilliant white façade; instead we see large advertisement billboards and a large sign heralding "San Luis Shopping." Nevertheless, the white façade typical of the rural hacienda and partly decorated with wrought-iron fences here maintains the visual coherence of the colonial theme. The surrounding parking lots, which are especially visible from the front side of the mall, once again reveal the function of the building. Here the preference for higher-income consumers and car owners from Quito and the valley is expressed, while the pedestrian zones are quite limited. The back side of the mall—directed southwards vis-à-vis the volcanoes, the former San Luis hacienda, and an extended park area—displays the colonial theme in a more explicit manner. Here, windows and wooden balconies covered with blooming flowers adorn the building. In this part of the mall the *campanario* (clock tower) is located—the most outstanding element of the religious hacienda architecture.

The interior of the shopping center also contains several stylistic signifiers that refer to the colonial motif. Arcades with round arches mark the entire interior, and bronze heads of cows and horses adorn the capitals of the columns. A tower with wrought-iron fences, which obscure the central elevator, is placed at the center of the mall. One of the most outstanding colonial features in the interior of the building is the *patio de comidas* (food court) on the third floor, which is dominated by dark wooden constructions and stairs. Like a theme park the San Luis Shopping Center creates both an imaginarium and an experience of the colonial in the Chillos Valley.

The myth of coloniality lies at the heart of the conflict-ridden, violent formation of Latin American societies; therefore it is convenient here to explore the connotations of coloniality. For the conceptualization of coloniality, one can rely on the recent debate in Latin American Postcolonial Studies (e.g. Moraña, Dussel, and Jaúregui 2008). Aníbal Quijano (2000) argues that alongside economic and political conquest, a "coloniality of power" based on identity politics has been established. In the classification of "racial differences," the European self is constructed as a norm, against which other groups are constructed as inferior in order to serve the needs of a racially stratified division (and exploitation) of labor. Other postcolonial critics such as Walter Mignolo (2000) and Álvaro García Linera (2004) have argued in the same way, contending that the colonial situation cannot be reduced to a mere historical period. On the contrary, it can be pointed out that the American societies are characterized by deep colonial structures that interfere with modern, republican layers. In terms of the racial divide and the exclusion of indigenous and African-descendant populations, coloniality continues to persist. In Ecuador, the indigenous movement has challenged this "power of coloniality." The movement's powerful protest demonstrations in the 1990s led to a constitutional redefinition of the nation state as a pluricultural entity and thus questioned white supremacy while recognizing multi- and interculturality and ethnic differences.

Although coloniality has its origins in the initial conquest, this does not mean that the ongoing power of coloniality should be conceptualized as a monolithic, never-changing structure. The postcolonial critic Mary Louise Pratt advises:

> If one seeks to establish the continuity across time of a "colonial legacy" one will fail to explain the processes by which this "legacy" has been and continues to be ongoingly renewed and integrated into a changing world through continuing permutations of its signifying powers, administrative practices, and forms of violence. (2008: 461)

Now, the key questions with regard to the San Luis Shopping Center concern how coloniality is reframed here, and to what extent the issues of exploitation, discrimination, and violation—inherent in coloniality—are addressed.

In a transnational context of recognizing indigenous rights, of multicultural policies and powerful indigenous movements, it is nearly impossible for the white elites to rely on an unquestioned model of racial hierarchy, as it was the case until the mid-twentieth century. To use coloniality strategically, the concept must be de-historicized. The connotations of the colonial *longue durée*, especially the aspects of racism, exploitation, and colonial shame, need to be eradicated in order to make coloniality a concept that sells and can be consumed. Consequently, the national press coverage of the San Luis Shopping Center and the declarations of the developers of the mall do not contain any critical remark on the historic hacienda as a system of exploitation, forced labor (*huasipungaje*), and political control.

This paradoxical form of erasing the signification and strategic use of coloniality is conceptualized here as retro-coloniality. In this fluid process of signification, which articulates different discourses, retro-coloniality can be understood as an empty signifier (Laclau and Mouffe 1985). While Jean Baudrillard (1995) claims that there is a tendency in postmodern societies whereby the architectural form has only a decorative function and becomes a simulacrum, I argue here that the form and the empty signifier "retro-coloniality" are related to strategic uses of identity.

One main aspect characterizing shopping centers is the narration of a story, which is related to dynamics of theming and imagineering, as the Disney company has developed them in US-American urban spaces (Gottdiener 1995, 2001). However, it is important to emphasize the fact that this story is not arbitrary, as developers and retailers depend on gaining their target groups' acceptance of the theming. In a cultural context marked by the crisis of the identitarian and socioeconomic positions of the middle and upper class in the Ecuadorian highlands, it is a promising marketing strategy to rely on emic or local cultural elements to create notions of authenticity, stability, and nostalgia. Therefore, the developers of the mall have argued that they take the specific "sense of place" of the Chillos Valley into account. The San Luis mall uses the cultural elements as a form of branding to distinguish itself from other malls in the area. While Dehler/Dekron, the commercial group that built the mall, relies upon Italian design for its shopping center San Marino in the coastal town of Guayaquil, in the more traditional highland capital Quito, it invokes with the Quicentro in the modern part of the city and with the San Luis Shopping Center on the outskirts a local sense of place and the nostalgic idea of the "good old" colonial and Hispanic past. As the administrator of the San Luis Shopping Center, Paola Padovani, highlights:

"Its architecture tries to rescue the façade of the typical houses of the Ecuadorian Sierra" (quoted in *Dinero*, August 30, 2006). This renewed importance of nostalgia in postmodernity (Hutcheon 1988, Boym 2002) can be interpreted as a reaction to the recent crisis in the process of identity politics. Vis-à-vis the dissolution, liquidation, and hybridization of identities (Bauman 2000, 2007) as well as the end of white supremacy, Western modernity, and development as hegemonic ideologies, colonial nostalgia can be regarded as an antagonistic dynamic of self-reflexive stabilization of identities in time and space. Nostalgia is thus not a simple return to the past, but a strategic positioning with regard to an uncertain future. This dynamic is highly important in the field of identity politics, as it implies an appropriation of the past that aims not to come to a critical discussion of history but instead aims to lead to a harmonization of the past by blending out social conflicts. The coloniality of power, with its patterns of exploitation and humiliation as well as its continuity in the present racism and the exclusion of the indigenous population, remains hidden.

The San Luis Shopping Center's evoking of a colonial past can be seen as a nostalgic "invention of tradition" (Hobsbawm and Ranger 1996). Although the Chillos Valley was the breadbasket of Quito in the colonial era and still contains several outstanding colonial landmarks such as haciendas and churches, the San Luis Hacienda, which lends the mall its name and was hailed in the national press as "one of the oldest haciendas" (Aguilar 2004: 40), was not built before the twentieth century. To sum up the paradox: The colonial architectural elements do not relate directly to colonialism as a historic period, but instead contribute to a symbolic retro-colonization of the urban landscape.

Thereby one can witness an articulation of "heritage" and "authenticity," on the one hand, and strategies of marketing and commodification, on the other, which lead to new forms of nostalgic consumer identities. While thinkers like Zygmunt Bauman (2007) and Jean Baudrillard (1998) have pointed out the overwhelming general significance of consumption for the process of identity formation in postmodern societies, dynamics such as the retro-waves (Guffey 2006) or the booming of heritage tourism (Porter 2008, Ashworth and Tunbridge 2000, Dallen and Boyd 2003) reveal the importance of the marketing and consumption of nostalgia. The concepts of "the pastime of past time" (Hutcheon 1988) or "histourism" (Römhild 1992) are concerned with the aestheticized consumption of easily "digestible" history, while the continuity of deep historical structures in contemporary societies is not taken into account.

The consumption of places echoes this consumption of time. John Urry has argued that places are subordinated to a tourist way of seeing, the "tourist gaze" (1990, 1992), which transforms "sites" into "sights." The commercial strategy to build a shopping center as site/sight with an authentic local character that corresponds to the heritage of the region finds its counterpart in municipal strategies for cultural and heritage tourism. Currently, the local government (*municipio*) strategically relies on the tradition and authenticity of the Chillos Valley to foster cultural tourism, using the fact that the area is a traditional leisure space for Quito's middle class. The San Luis Shopping Center counters multiple small popular restaurants, swimming pools, and other facilities in the valley. Similarly, the municipality of Rumiñahui, with the help of the Ministry for Tourism, wants to declare the mall a tourist destination (*Universo*, September 2, 2006). The self-reflexive aspect of the suspension of disbelief about the themed character of the mall may be illustrated in the interview we conducted with the Rumiñahui municipality's Director of Planning, Jorge Sosa. On the one hand, he highlights the importance of the mall in the municipality's strategy to foster heritage tourism, but, on the other hand, he also sees the constructed historic theme of the mall: "It is a postmodern architecture that they put us there, and thus it reflects this duality of the new and the old" (2008).

The mall's functional use in daily life determines its local perception, as people tend to mostly disbelieve the colonial theming of San Luis Shopping. Here one cannot detect any desire to decode the mall in the preferred reading, which is explicitly expressed in many conversations we had with the neighbors of the mall in the Chillos Valley. With a growing cultural and spatial distance, there seems to exit a greater wish to believe in the theming.

Post-Rural Pleasures

The San Luis Shopping takes up the colonial theming—usually referred to the urban colonial centers in the Americas—and connects it with notions of rurality. This practice has a strong imaginary force, as the motif of the colonial hacienda is one of the most important icons of the rural highlands in Ecuador. Several features of the exterior and interior of the mall create the countryside theme. The interior of the shopping center rounds off its enunciation of the rural imaginary by creating a "natural" ambiance with trees and bushes as well as with decorations made of bougainvilleas. On the exterior, especially the southern (back) side of the mall presents itself as a natural environment with fountains, palms, green areas, playgrounds for children, stone benches and tables, and the Avenida San Luis named after the Shopping Center. These elements evoke a colonial past and a naturalized, idyllic atmosphere, but they also serve functional goals: the backside of the mall can be seen as its retail area, designed to stimulate the consumption of food and drink. To create a "real" and "authentic" rural atmosphere, the shopping center hosts the *Café de la Vaca*, one of the most famous, hacienda-like restaurants in the rural area and originally situated on the highway south of Quito.

This connection between notions of rurality and the consumption of food and drink is best expressed in the *patio de comidas* (food court) inside the mall. It evokes the rural imagination of a trip to the hacienda, which is deeply rooted in the myth of country life in the Ecuadorian highlands. The social anthropologist Maria Dabringer (2011) argues that in Quito, the typical food can be seen as an important part of a complex system of references that creates traditions and stability. Although the *patio de comidas* is styled like many rural restaurants, often located in former haciendas, which are frequented by the urban middle class on weekend excursions, the offered food does not differ from that of other malls, serving pizza, French fries, and even sushi. Nevertheless, public acceptance seems to be broad, leading one interviewee from the Rumiñahui municipality to state in an interview—off the record—that the *patio de comidas* put an end to many traditional restaurants of the area.

This use of the myth of rurality can be explained with the concept of the "post-rural" (Hopkins 1998: 77). The post-rural refers first to the aestheticization or symbolization of the material countryside into a "postmodern good" by virtue of its commodified sign value (Lash and Urry 1994). As the rural is connoted with aspects of eternity and the natural, it offers—as retro-coloniality does—a stabilizing element for the white middle class identities in crisis. As the rural implies domesticated nature—including the indigenous "naturals," as they were called in colonial times—it expresses the will to control and superiority. Furthermore, through the connotations of relaxation and good food, it can be related to important elements of postmodern consumer identities.

The latter aspects demonstrate that the post-rural is not only a merchandizing trick, but that it also appeals to the sentiments and wishes of the target consumers. It thus exists as a reflexive attitude towards post-rurality; although the consumers know about the constructed character, they nevertheless *want* to believe in its authenticity. The reflexivity leads to "uneasy pleasures," a "tension between the consumers' willing suspension of disbelief and their knowledge of an advertiser's [or shopping center's] persuasive intentions" (Hopkins 1998: 77). This perception also depends upon identitarian positionings and loci of enunciation. While many consumers from Quito want to perceive the mall as an integral part of a rural landscape, this is not the case for the neighbors of the mall. Rafael Z., a villager of the valley who worked decades ago in the former San Luis hacienda, answered to the question "[a]nd the house of the hacienda, is it like the structure of the mall or does it have nothing to do with it?" by saying "[n]o, nothing to do, nothing to do" (2008). He later goes into detail about the rural landscape of the hacienda: "Nnnno, imagine exchanging a stretch [of plants] for so many square meters of concrete. Something that really had trees, grass, that was something natural, old trees that were—as my Dad had told me—more than 200 years old."

New Urbanity: Constructing *Heimat*

The exterior architectural complex—together with the corresponding changes in the urban transportation systems—obviously leads to a new and accelerated process of resignification of the suburban space of the valley. Conversely to that, it is also the interior of the mall that provokes changes in the urban spaces. Although the interior structure is hidden from the external perception of landscape, it can be understood as a feature that forms part of the urban landscape. The mall's interior is a hybrid public–private space—a space that satisfies public needs but is controlled

Figure 5.2   *Patio de Comidas* **of the San Luis Shopping Center. Photograph**
by the author

by a private enterprise and private security guards—that simulates an urban
landscape. Although this landscape cannot be perceived by the eye of an external
observer, it forms part of the people's cognitive maps and urban imaginaries.

The San Luis Shopping Center consists of three storeys: one central hall on
the first floor, two lateral corridors on the second floor, and two smaller areas on
the third floor. As we have seen, one finds signifiers of the colonial and the rural
on all three floors. Although the rural colonial hacienda motif dominates, it is not
the only theme of the San Luis Shopping Center. Like most other malls, San Luis
Shopping also creates a feeling of urbanity. The interior of the mall represents
an exterior space that evokes a nostalgic image of a European town of the turn
of the twentieth century. The corridors resemble streets illuminated by wrought-
iron lanterns. Small squares with benches and kiosks subdivide the corridors. On
the first floor of the mall one finds replicas of art booths similar to those set up in
Quito's Ejido Park on weekends, which appeal to the urban middle class as well

Figure 5.3   Interior of the San Luis Shopping Center. Photograph by the author

as to domestic and international tourists. As in many other shopping centers, San Luis Shopping simulates the European city, creating not only a proper atmosphere for consumption, but also the cultural experience of *flâneur*-like shopping. A roof constructed of steel and glass and abundant windows on the third floor provide natural light. These features strongly recall European Historicist architecture of the turn of the twentieth century such as English Victorian Gothic, the German Wilhelminian or French Empire Style and their blending of industrial, neomedieval, and neoclassical elements, as well as a resemblance to the arcades.

Despite these explicit features which characterize the form of the building, San Luis Shopping features a host of functional elements typical for shopping centers in general (Goss 1993). A broad escalator leads into the mall. To the right hand-side one sees the anchor store, which belongs to the supermarket chain Supermaxi, whereas on the left side of the entrance hall one passes by a branch of the most important local bank, Banco de Pichincha. An eye-catcher—a clock tower with a wrought-iron elevator inside, surrounded by a fountain and stone stairs placed at the center of the hall—soon attracts the viewer's gaze. A paradoxical "timeless time" characterizes the wrought-iron clock, as it features a large face, but its hands are so tiny that is impossible to read the time. Changing events such as tombolas for luxury cars or the exposition of a gigantic Christmas tree also take place at the center of the hall. While there are stone staircases and the slow elevator at the center of the mall, the faster-moving escalators are located at the far ends of the building, thus forcing consumers to pass by the entire gallery of stores in order to reach them. The restrooms are located on the third floor, preventing passers-by from using the facilities without first passing the stores.

Although there exists a planned arrangement of the perception and use of the mall, consumers' appropriations of the San Luis Shopping Center depend on the specific sociocultural position of the consumer. Carlos Mancheno, administrator of the store Pinto in the mall, observes that the building's "semirural location makes it a center in which various socioeconomic segments are interested in. The lowest sector of the scale will visit the mall to go for a walk, ... while the middle and upper sectors of the scale will make use of its entire spectrum [of possible purposes]" (2008). Our own observations in the mall confirm Mancheno's perception. The youth of the nearby *Escuela Politécnica del Éjército* (Army Polytechnic School) and the La Salle High School have declared the San Luis Shopping Center to be their new meeting place, thus changing the juvenile socio-geographic landscape from the *Triángulo*, an open public place, and the older River Mall to the San Luis Shopping Center. Their preference highlights the social importance of the mall as a meeting place, as explained by one student whom we interviewed in the mall: "Then you go to the San Luis ..., brother, you walk around with all the others, you meet somebody, you meet another one to chat with, and like this you make yourself, brother. Your drink your beer and you go" (Student 2008). The local youth culture does not take the retro-colonial architecture into account, and the mall is in no way seen as an authentic local place. Instead, it appears as the representation of a postmodern globalized "non-place" (Augé 1995), that connects the valley with the patterns of global consumer culture.

The creation of an urban atmosphere is a feature that can be found in nearly all shopping centers. The main urbanist purpose of the ideal shopping center, as it was conceived by the Austrian urban planer Victor Gruen in the mid-twentieth century was to compensate for the lack of a historic city center in the US-American city by creating a city center atmosphere following the model of European towns. As we have seen, this also applies to the San Luis Shopping Center. However, with regard to the creation of urbanity, the San Luis Shopping Center goes even further: not only can one see the tendency to create a quasi-urban center within the mall, but the mall itself serves as a reference point from which to recenter the suburban spatial structure in the Chillos Valley.

While the past decades, particularly in the Americas, have been characterized by enormous rates in urban growth and, since the 1970s, by urban sprawl and the decay of city center districts and downtown areas, one is currently witnessing a "return to the center" (Rojas 2004, Herzog 2006). This entails the reoccupation of urban space in the center districts by hegemonic groups and often finds its expression in processes of gentrification, urban renewal, and slum clearance as well as in new urban consumer and lifestyle identities. This return to the center is evident in the process of urban renewal in the historic city center of Quito as well as in the suburban project of the Shopping Center. At the beginning of the twenty-first century the Chillos Valley had a population of 250,000 inhabitants and a high growth rate, making an estimated 400,000 inhabitants to be projected for 2010 (Escobar 2006: 43). In the context of a lack of urban planning, the private enterprise faces the challenge of creating a city center in the dispersed spatial structure of the valley. This planning includes not only the San Luis Shopping Center with its entertainment facilities, especially the ten cinemas, but also a hospital—the medical sector is an emerging market in Ecuador—a residential zone, a town center, and a church (Interview with Carmen Recalde 2008). Paola Padovani of DK Management, the administrators of the mall, explains: "we hope to open the Hospital San Luis that will work in alliance with the *Hospital Metropolitano* and a town center that will account, alongside offices, for the economic development of the zone" (*Dinero*, Augsut 30, 2006). Therefore, developments would necessitate changing even the routing and configuration of streets and roads.

These features are highly functional, in the sense of the central place theory of Walter Christaller and are urgently needed in the valley. Nevertheless, questions arise relating to the privatization of services and other policies motivated by the aim of maximizing profits. Such concerns are not only problematic for the health sector, but are also particularly relevant in terms of citizenship, for example, when a private enterprise runs a town center. For the middle- and upper-class sectors who can afford to buy these services, the standard of living in the valley rises, which in turn leads to an accelerated process of gentrification. The question remains open as to whether the lower classes—often of indigenous backgrounds—stand to benefit from the new suburban mega-project. These questions of inclusion and exclusion are also expressed in regional identity politics. In contrast to urban tendencies of individualization and anonymity, the new center is constructed as a familiar small town following the principles of New Urbanism. In this sense, the administrator, Carmen Recalde, understands the mall in terms of a "new home" (2008) for the population of the valley, thus creating *Heimat*.

Fractal Geopolitics of Identity

The empirical visual analysis shows that the San Luis Shopping Center consists of different stylistic elements such as the colonial, the modern, the natural, the rural, and the European city, as well as of diverse functional aspects. In so doing, it combines a new allusion to cultural heritage and nostalgia with the ideal typical blueprint of a shopping center. Following the sociosemiotic approach, one can argue that the façade and the colonial theming are a form of expression that covers up the functional purpose of the mall, the substance of the content which is the aim to make profit. Mark Gottdiener argues:

> In the case of the malls, they can be understood best as the intersection sites of
> two distinct structural principles. On the one hand, the mall is the materialization
> of the retailer's intention to sell consumer goods at a high volume under present-

day relations of production and distribution. The mall, then, is a "substance of the expression" engineered for the realization of capital in a consumer society. As such, it embodies particular design artifacts—i.e., the morphological elements, or the "form of expression"—which are instrumentally designed to promote purchasing. (1995: 84)

This ideological gap between form and function, content and expression, is the secret of the "magic of the mall" (Goss 1993) in the sense of the sort of "magic" found in vaudeville theater, that is, magic which directs the attention of the spectator. The colonial discourse about the Shopping Center propagated in the print media as well as by the administration and the developers of the mall reinforces this visual impression.

Nevertheless, in the case of the San Luis Shopping Center, I argue that the "form of expression" has deeper ideological implications than the stimulation of the desire to consume. Furthermore, the form of expression is here related to the myth of coloniality, thus representing a forceful intervention in the field of identity politics. In order to work out these dimensions, it thus seems important to recontextualize the mall with regard to broader social processes such as identity politics in Ecuador and urban renewal in Quito.

From the Chicago School to Pierre Bourdieu, there exists a strong research tradition that seeks to relate social processes and structures to physical space. Although a homology between spaces and identities certainly exists, it may be too simple to speak of a complete determination of space by social structures. In concrete cases, the articulation between certain groups and their appropriations of space is related to an entire host of different actors, to the imagination of landscapes, to the proper logics and effects of space, and to the social context.

In Quito, the white urban middle and upper classes have appropriated the Chillos Valley and, since the 1990s, also the historic city center. Currently, it seems that the construction of the San Luis Shopping Center expresses this symbolic appropriation of space through elements of retro-coloniality and social distinction. With regard to the social structure of Ecuadorian society, retro-coloniality and the post-rural are strongly linked to the middle and upper classes in the highlands, which consist of a fusion of former land-owners, the state bureaucracy of the 1970s, and, since the 1990s, an emerging financial sector. As in other Latin American countries (Argentina for instance), the allusion to a rural, aristocratic imaginary serves as a means of distinction from other emerging social strata and as a form of "cultural mimesis" to the lifestyle of the traditional rural elites (Svampa 2008). In the case of identity politics in Ecuador, retro-coloniality may be seen as both a mode of social and cultural distinction from the indigenous population and a way to redefine the principles of vision and division of the social world in Ecuador.

This process of retro-colonizing the landscape is not an isolated phenomenon. It can be argued that the construction of the San Luis Shopping Center links the Chillos Valley aesthetically and symbolically to the urban renewal of Quito's historic center. In the Ecuadorian capital, the most complete urban colonial ensemble in Latin America and a UNESCO World Cultural Heritage site since 1978, retro-coloniality is of particular importance. The principle finds expression in programs and projects of urban renewal and in the restoration of the historic city center, conducted under a strong presence of transnational actors such as UNESCO or the Inter-American Development Bank. In the 1990s in particular, the historic city center became a magnet for national and international heritage tourism, using retro-coloniality to brand and market the city. In a kind of heritage mimesis, the developers of the San Luis Shopping Center follow the development path employed by both urban renewal processes and the reawaking of cultural heritage in the city center. With its multiple references to colonial architecture and aesthetics, the mall can be seen as echoing the reinforced process of urban revitalization of the historic "colonial" center of Quito.

Many urban sociologists and geographers have argued that the Latin American city is characterized by the emergence of new fragmented spaces, which take the form of gated communities, shopping centers, and other enclosed structures—an urban process that Peter Marcuse analyzes by employing the concept of citadel. Marcuse defines the citadel as a "spatially concentrated area in which members of a particular group, defined by its position of superiority, in power, wealth, or status, in relation to its neighbors, congregate as a means of protecting or enhancing that position" (1997: 247). With regard to cultural politics, these spaces are mostly characterized by a lifestyle-related theming. However, I maintain that the notion of isolation and enclosure in a "spatially concentrated area" where privileged groups "congregate as a means of protecting or enhancing" is misleading in the present case. First, the concept of the citadel evokes a notion of defense. In the Ecuadorian case, one can instead observe a logic of identitarian expansion and appropriation of central places that are linked to an integrated urban landscape. Second, I argue that, from the perspective of an upper middle-class consumer, these spaces are not isolated, but linked: functionally by highways and imaginatively by the production of a retro-colonized landscape. Arguing with the models of perception, the iconic spatial elements serve as symbolic landmarks that provide the landscape with new meaning. In this sense, one cannot speak of "insular spatial formations" (Marcuse 1997: 249), but instead must speak of an urban landscape in the form of a fractal archipelago. In the context of the retro-colonialization of the urban landscape, the San Luis Shopping Center can be conceived as an iteration, a self-similar reproduction, of the colonial city center. Thereby two different spaces are articulated in terms of a fractal archipelago that is framed by the urban imaginary of a retro-colonialized landscape.[4]

This new landscape is related to the social context of the rising importance of multicultural identity politics of the Ecuadorian state. In line with the German philosopher Wolfgang Welsch, it can be argued that multiculturality

---

4   For a theoretical discussion of fractal dynamics in the field of identity politics in the Americas see Thies and Kaltmeier 2009.

takes up the problems which different cultures have living together *within one society*. But therewith the concept basically remains in the duct of the traditional understanding of culture; it proceeds from the existence of clearly distinguished, in themselves homogeneous cultures—the only difference now being that these differences exist within one and the same state community. (1999: 196)

This statement reflects in itself the aforementioned fractal archipelago of identitarian spaces. Thus, it becomes important to mention that the white middle class identity politics, with its resemblance to Hispanism and retro-coloniality, represents a reaction to the powerful protests of the indigenous movements. The distinction from indigenous popular groups and a reemergence of coloniality and consumerism—the possibility to consume as common element of the fragmented middle and upper classes—substitute for the loss of an inherent feature of cultural cohesion after the loss of development and progress as integrating ideology. While the 1990s were characterized by ethno-politics and ethno-development, it seems that the upper and middle classes in Quito account for a new ethno-consumption in a "mall-ticultural" society.

Works Cited

Aguilar, C. 2004. Ciudades en miniatura. *Vistazo*, October 1.

Aitken, S. and Craine, J. 2005. Visual methodologies: what you see is not always what you get, in *Methods in Human Geography,* edited by R. Flowerdew and D. Martin. Harlow: Prentice Hall, 250–69.

Augé, M. 1995. *Non-Places: Introduction to an Anthropology of Supermodernity*. London: Blackwell.

Baudrillard, J. 1995. *Simulacra and Simulation*. Michigan: University of Michigan Press.

Baudrillard, J. 1998. *The Consumer Society: Myths and Structures*. London: Sage.

Bauman, Z. 2000. *Liquid Modernity*. Cambridge: Polity Press.

Bauman, Z. 2007. *Consuming Life*. Cambridge: Polity Press.

Bourdieu, P. 1984. *Distinction: a Social Critique of the Judgment of Taste*. Cambridge: Havard University Press.

Bourdieu, P and Wacquant, L. 1992. *Invitation to Reflexive Sociology*. Chicago.

Boym, S. 2002. *Future of Nostalgia*. New York: Basic Books.

Büschges, C. and Pfaff-Czarnecka, J. (eds.) 2007. *Die Ethnisierung des Politischen: Identitätspolitiken in Lateinamerika, Asien und den USA*. Frankfurt/M.: Campus.

Carrión, F. (ed.) 2001. *Centros históricos de América Latina y el Caribe*. Quito: FLACSO.

Carrión, F. 2005. Los centros históricos en la era digital en América Latina, in *Ciudades translocales: espacios, flujos, representación,* edited by R. Reguillo and M. Godoy. Guadalajara: ITESO, 85–108.

Carrión, F. and Hanley, L. (eds.) 2005. *Regeneración y revitalización urbana en las Américas: hacia un estado estable.* Quito: FLACSO.

Centro Comercial Hacienda Santa Bárbara (n.d.). *Historia* [Online: Centro Comercial Hacienda Santa Bárbara]. Available at: http://www.haciendasantabarbara.com.co/historia.htm [accessed: June 4, 2010].

Cosgrove, D. 1998. *Social Formation and Symbolic Landscape.* 2nd Edition. Madison: University of Wisconsin Press.

Dabringer, M. 2011. Consumo local/global en Quito: "tradiciones globalizadas" en el contexto urbano, in *Culturas Políticas en los Andes,* edited by C.Büschges, O.Kaltmeier, S. Thies, and P. Birle. Frankfurt/M.: Vervuert

Dallen, T. and Boyd, S. 2003. *Heritage Tourism.* Harlow: Prentice Hall.

Dann, G. 1998. There's no business like old business: tourism, the nostalgia business of the future, in *Global Tourism,* edited by W. F. Theobald. 2nd Edition. Oxford: Butterworth-Heineman, 29–43.

Debord, G. 1995 [1967]. *The Society of the Spectacle,* New York: Zone Books.

Duncan, J. 1990. *The City as Text: The Politics of Landscape Interpretation in the Kandyan Kingdom.* Cambridge: Cambridge University Press.

Ellin, N. 1997. *Architecture of Fear.* Princeton University Press.

Escobar, M. 2006. El Shopping del Valle. *Vistazo,* October 5.

García Canclini, N. 1995. *Consumidores y Ciudadanos.* México, D.F.: Grijalbo.

García Linera, Á. 2006. State crisis and popular power. *New Left Review,* 37, 73–85.

Goss, J. 1993. The "magic of the mall:" an analysis of form, function, and meaning in the contemporary retail built environment. *Annals of the Association of American Geographers,* 83(1), 18–47.

Gottdiener, M. 1986. Culture, ideology, and the sign of the city, in *The City and the Sign,* edited by M. Gottdiener and A. Lagopoulos. New York: Columbia University Press, 202–18.

Gottdiener, M. 1995. *Postmodern Semiotics: Material Culture and the Forms of Postmodern Life.* Cambridge: Blackwell.

Gottdiener, M. 2001. *The Theming of America.* Boulder: Westview Press.

Graham, B. and Howard, P. (eds.) 2008. *The Ashgate Research Companion to Heritage and Identity.* Farnham: Ashgate.

Gregory, D. 1994. *Geographical Imaginations.* Cambridge: Blackwell.

Guffey, E. 2006. *Retro: The Culture of Revival.* London: Reaktion Books.

Hall, S. 1997. *Representation. Cultural Representations and Signifying Practices.* London: Sage.

Herzog, L. 2006. *Return to the Center: Culture, Public Space, and City Building in a Global Era.* Austin: University of Texas Press.

Hobsbawm, E. 1996 [1983]. Introduction: inventing traditions, in *The Invention of Tradition,* edited by E. Hobsbawm and T. Ranger. Cambridge and New York: Routledge, 1–14.

Hopkins, J. 1998. Signs of the post-rural: marketing myths of a symbolic countryside. *Geografiska Annaler B*, 80(2), 65–82.

Hutcheon, L. 1988. *A Poetics of Postmodernism: History, Theory, Fiction.* New York and London: Routledge.

ICSC 2009. *2008 Winners: ICSC 32nd Design and International Development Award.* [Online: International Council of Shopping Centers]. Available at: http://www.icsc.org/designawards/2008_DIDA_awards_winners.pdf [accessed: March 5, 2009].

Kaltmeier, O. 2007. ¿Politización de lo étnico y/o etnización de lo político? El espacio político en el Ecuador en los años noventa, in *Poder y etnicidad en los países andinos*, edited by C. Büschges, G. Bustos and O. Kaltmeier. Quito: Corporación Editora Nacional, 195–216.

Kaltmeier, O. 2011. Hacienda, Staat und indigene Gemeinschaften: Kolonialität und politisch-kulturelle Grenzverschiebungen von der Unabhängigkeit bis in die Gegenwart, in *Der verweigerte Sozialvertrag: Politische Partizipation und blockierte soziale Teilhabe in Lateinamerika*, edited by H.-J. Burchardt and I. Wehr. Baden-Baden: Nomos.

Koolhaas, R. 1997. Bigness or the problem of large, in: *S, M, L, XL*, by O.M.A., R Koolhaas, and B. Mau. 2nd Edition. New York: Monacelli Press, 494–517.

Laclau, E. and Mouffe, C. 1985. *Hegemony and Socialist Strategy: Towards a Radical Democratic Politics.* London: Verso.

Lash, S. and Urry, J. 1994. *Economies of Signs and Space.* London, Thousand Oaks and New Delhi: Sage.

Marcuse, P. 1997. The enclave, the citadel, and the ghetto: what has changed in the post-Fordist US city. *Urban Affairs Review,* 33(2), 228–64.

Mignolo, W. D. 2000. *Local Histories / Global Designs.* Princeton: Princeton University Press.

Moraña, M., Dussel, E. and Jáuregui, C. (eds.) 2008. *Coloniality at Large.* Durham, NC: Duke University Press.

Pratt, M. L. 2008. In the neocolony: destiny, destination, and the traffic in meaning, in *Coloniality at Large*, edited by M. Moraña, E. Dussel and C. Jáuregui. Durham, NC: Duke University Press, 459–78.

Quijano, A. 2000. Colonialidad del poder, eurocentrismo y América Latina, in *Colonialidad del Saber, Eurocentrismo y Ciencias Sociales*, edited by E. Lander. Buenos Aires: CLACSO, 342–86.

Quintero, R. 1986. El estado terrateniente del Ecuador (1809–1895), in *Estado y nación en los Andes*, vol. 2, edited by J.-P. Deler and Y. Saint-Geours. Lima: Instituto de Estudios Peruanos, 399–406..

Reguillo, R. and Godoy, M. (eds.) 2005. *Ciudades translocales: espacios, flujos, representación.* Guadalajara: ITESO.

Rojas, E. 2004. *Volver al centro: la recuperación de áreas urbanas centrales.* Washington, DC: Inter-American Development Bank.

Römhild, R. 1992. Histourismus: zur Kritik der Idyllisierung, in *Reisen und Alltag: Beiträge zur kulturwissenschaftliche Tourismusforschung,* edited by D. Kramer and R. Lutz. Frankfurt/M.: Universität Frankfurt, Institut für Kulturanthropologie, 121–30.

San Luis eleva la oferta de malls en Quito. 2006. *El Universo,* September 2.

San Luis une a tres grupos económicos. 2004. *Dinero, Diario de Negocios,* October 5.

Svampa, M. 2008. Kontinuitäten und Brüche in den herrschenden Sektoren, in *Sozialstrukturen in Lateinamerika: Ein Überblick,* edited by D Boris et al. Wiesbaden: VS-Verlag, 45–71.

Thies, S. and Kaltmeier, O. 2009. From the flap of a butterfly's wing in Brazil to a tornado in Texas? Approaching the field of identity politics and its fractal topography, in *E Pluribus Unum? National and Transnational Identities in the Americas / Identidades nacionales y transnacionales en las Américas,* edited by S. Thies and J. Raab. Münster and Tempe, AZ: Lit Verlag and Bilingual Review Press, 25–46.

Una inversión muy esperada … . 2006. *Dinero, Diario de Negocios,* August 30.

Urry, J. 1992. The tourist gaze revisited. *American Behavioral Scientist,* 36(2), 172–86.

Urry, J. 1990. *The Tourist Gaze: Leisure and Travel in Contemporary Societies.* London: Sage.

Welsch, W. 1999. Transculturality—the puzzling form of cultures today, in *Spaces of Culture: City, Nation, World,* edited by M. Featherstone and S. Lash, London: Sage, 194–213.

Yúdice, G. 2003. *The Expediency of Culture.* Durham, NC: Duke University Press.

Zukin, S. 1995. *The Cultures of Cities.* Malden, MA, and Oxford: Blackwell.

Interviews

Mancheno, C. Almacenes Pinto, interviewed by A. Schlenker and A. M. Egas, January 2008.

Recalde, C. Administrating Manager of the San Luis Shopping Center, Email-interviewed by A. Schlenker and A. M. Egas, January 31, 2008

Sosa, J. Director of Planification of the Rumiñahui Municipality, interviewed by A. Schlenker and A. M. Egas, January 2008.

Student, 16 years. Interviewed by C. Ahassi, January 2008.

Z., Rafael. Neighbor of the San Luis Shopping Center, interviewed by D. Lasso, January 2008.

Chapter 6

Religion and Culture Set in Stone:
A Case Study of the Jewish Community
Center of Metro Detroit

Julie TelRav

This chapter will attempt to contribute to a better understanding of the reciprocal relationship between architecture and the development of religious and cultural community. It will do this by examining how the Detroit Jewish community seeks to maintain its historic ties to religious and cultural tradition, but also to be innovative and make Judaism more attractive and relevant to a declining population, through its adaptation of two architectural environments—the buildings of the Jewish Community Center of Metropolitan Detroit (JCC). This is a comparative case study of the different ways that architecture is used to facilitate social interaction and group bonding, enact imaginations and desires for community, and foster active planning for future social change.

The JCC makes for an especially fruitful case study because it operates two buildings in two very different suburban neighborhoods of the greater Metro Detroit, Michigan, area: the Kahn building in West Bloomfield and the Jimmy Prentis Morris (JPM) building in Oak Park. Both offer programming and services to its constituents but in very different environments. The JCC's explicit mission is to "support Jewish unity, ensure Jewish continuity, and enrich Jewish life"— or, to be a primary locus of Jewish identification. Therefore, both buildings must accommodate traditional Jewish observance while still remaining a non-denominational[1] communal gathering place and a broad representation of the history, tradition, and ethics of the Jewish community to the city of Detroit.

Broadly stated, my objectives in researching these institutions were threefold. I wanted to better understand 1) the impact architecture has on social behavior and group interactions within a communal religious institution, 2) how physical environments can support or inhibit expressions of cultural and/or religious identity and inform searches for meaning in community members, and 3) how the changing needs and goals of the community dictate design strategies and guide adaptation of the existing environment, particularly in the current context of place-

---

1   Non-denominational Jewish institutions such as the JCC of Metro Detroit do not affiliate with any of the commonly recognized branches of the Jewish community, such as Orthodox, Conservative, Reform, or Reconstructionist Judaism.

based marketing strategies. More specifically in this article, I sought to understand the role of the JCC architecture in attracting and sustaining a declining Jewish community. In what ways do the buildings make people feel comfortable, promote communal affiliation among Jews, or foster relationships between Jews and the non-Jewish community, especially where synagogues and other Jewish institutions cannot? How do the buildings provide space for meaningful participation in Jewish life and ritual while still recognizing religious and cultural diversity? How does the Center utilize its designed physical spaces to help promote and differentiate itself and convince local Jews that both the JCC and modern-day Judaism have continuing relevance and value for the community?

Generally speaking, by "architecture" I am referring to those inhabitable objects brought into being through three interconnected processes. First is the creative design process that is used to communicate ideas through elements of form, scale, symbolism, and sensory stimulation. The second includes the socioeconomic processes which help define the need for the architectural object, and whose financial and social resources help bring it into being. And third is the ongoing process of interpretation and modification by its users. What also makes architecture sociologically distinctive is that it includes the less easily definable issues of "place," "space," and "site." These, beyond architecture, are more than something physical. Places, spaces, and sites can also provide a cognitive framework, as well as a form of social organization, in which actors relate to one another, events occur, and symbols are organized, presented, and received.

This project was informed by the Production of Culture theoretical approach, which is interested in the mechanisms through which cultural symbols are produced, reproduced, and manipulated in an effort to shape the beliefs, values and, consequently, the behaviors of people and groups within society. This occurs either by the slow and often unconscious, process of accretion or through the deliberate creation and manipulation of symbols in order to affect the actors who engage with them (Peterson 1979). In particular, Magali Sarfatti Larson (1997) explains how architecture, as a distinctive type of cultural artifact, has two essential and deliberate characteristics: it both denotes and connotes symbolic meanings and references. Denoted references suggest functional possibility as well as cultural and historical typological codes. While denoted references draw upon the slow accretion of symbolic imagery, connoted meanings are generally departures from our expectations of denoted meanings—a deliberate production and manipulation of cultural symbolism. They call our attention to another kind of symbolic script, one of messages and shared significance that "moves the users from a *form to an ideology* of inhabitation" (68, italics in the original). Also, the work of Robert Wuthnow begins to show the links between the production of architecture and that of modern day spirituality. In his book *Producing the Sacred*, he takes a controversial approach, claiming that the realm of the "sacred," and public religion in particular, is deliberately produced and that purveyors of modern religion, much like architects and designers, manipulate their adherents through the conscious and creative use of symbols. As he says, "public religion is produced—it is the

Figure 6.1   JPM Building, Oak Park, IL. Photograph by the author

product of a complex system of organizations that expend resources to bring the sacred into a relationship with the social environment" (1994: 151).

The architecture of the JCC of Metro Detroit was created in response to complex religious, political, social, and economic factors. Thus, its buildings are rich in both practical denoted references as well as the more symbolic connoted references. Both JCC campuses have their own distinctive design and are located in suburbs considered part of the Greater Detroit Metropolitan Area. The JPM building was originally built in the northwest suburb of Oak Park in 1956 as a satellite facility for Detroit's Dexter/Davison JCC. It was meant to accommodate a slowly shifting population, but in 1967 the Detroit riots caused a mass exodus of the Jewish population, many to the towns of Oak Park, Southfield, and Huntington Woods, making JPM the most accessible facility for the majority of Detroit's Jews.

By the late 1960s, the Jewish population in the northwest suburbs had expanded significantly, yet community planners assumed that the same socioeconomic forces that caused Jews to move out to these suburbs would only continue to push the Jewish population further north and west, leaving behind only elderly Jews and those with few other options. As such, they made little economic investment in JPM, assuming it would eventually close. But by the mid-1980s the Detroit Jewish leadership developed a new strategic plan. Noting the significant investment already made in Jewish communal infrastructure in the area, they proposed that it might be wiser to revitalize the community rather than continue chasing its fleeing members and rebuilding Jewish institutions. They developed a multi-pronged approach to neighborhood preservation that included restoration of Jewish infrastructure, such as the JPM facility. At that time 1.5 million dollars was spent to expand JPM into a full-service building and pay for necessary improvements. The money was used to add a swimming pool, fitness center, additional meeting rooms and new administrative offices (JPM Strategic Plan 2007).

Today, the Center is a 60,000 square-foot building that sits on 5½ acres. In addition to the aforementioned it has a gymnasium, a renovated social hall, and a kosher restaurant. The outdoor facilities include a community-built playground for elementary school-aged children, an updated toddler playground, and sports fields.

These days, Oak Park and Southfield would be described as diverse lower middleincome communities, while Huntington Woods attracts upwardly mobile, upper middle income families. All three have thriving Jewish populations; in fact the area is home to almost one third of Metro Detroit's 72,000 Jews. JPM has not been widely adopted by the Huntington Woods Jews but has developed a reputation as the symbolic center of the Oak Park/Southfield Jewish community. This is thought to be due to its commitment to economic and religious diversity. As an affordable and non-denominational institution, it is perceived as one of the only neutral spaces in the community where Jews of varied backgrounds, beliefs, and practices can come together (Magidson 2008).

However, even with the increased investment, the JPM facility is not on par with the newer Kahn building in scale or amenities. Like JPM, Kahn was built to accommodate the shifting location and needs of the Detroit Jewish community.

Figure 6.2　Kahn Building, West Bloomfield, MI

As wealthier Jews continued to move northwest into the townships of West Bloomfield and Farmington Hills, demand increased for a closer, newer, and more comprehensive Jewish leisure facility with functional space for a wide range of programming. Built in 1975, the Kahn building in West Bloomfield was created to be, and still is, the largest JCC in the country, probably in the world. At 330,000 square feet, it is more than five times the size of JPM. The façade of the Kahn building gives only the subtlest hint of what has been created on the inside. The exterior of the building avoids cultural or religious distinctiveness, representing itself instead as one of the many large institutions that make up the suburban landscape; it could just as easily be a corporate or civic institution as a Jewish communal building. From the outside it is a rather nondescript red brick building that has large characterless block forms and dark windows. Only its sheer massive size is impressive in that it is a symbolic testament to what organized philanthropy and communal ethnic commitment can create.

Yet while the outside aims to blend in with the local architecture, the inside markets itself to a discriminating religious and ethnic population. Not only does it seek to create a strong Jewish environment, but by catering to its wealthier clientele, the Center staff has attempted to create a place that would even attract Jews with the financial resources to make other (likely secular) choices for their recreational needs. They have repeatedly hired architects and designers to create lavish banquet and dining facilities, a state of the art fitness center, and outstanding museums. Throughout much of the public and event spaces, Kahn uses high end finishes, such as rich textiles, dark woods, elegant lighting, and granite and metal surfaces, giving the Center an elegant and sophisticated atmosphere. Today, the facility houses administrative offices, two gymnasiums, a 125,000 square-foot fitness center, an inline hockey center, three swimming pools, a library, a preschool, four museum and gallery spaces, a teen center, a performing arts theater, a large social hall, and a new kosher pizza shop. In addition, 40,000 square feet on the upper level have been rented to a Jewish day school. The facility sits on approximately 130 acres of green space in what has long been regarded as one of the most affluent townships in the country.

Social Interaction

Anne-Marie Fortier, who researched Italian Catholic community centers in Great Britain, argues that "we can perhaps think of community as a site that is constructed through the *desire for* community, rather than a site that fulfills and resolves that desire" (2006: 73, italics in the original). In many ways, this seems to be an apt description of the Kahn building as well. In the words of the first president of the JCC, Hugh Greenberg, Kahn was conceptualized as a "totally self-contained Jewish environment in which to perpetuate [the community's] traditions and … forge its future" (1975). Indeed, Kahn was built to be the most impressive facility of its kind in the United States, and the immense scale and amenities of the facility allow practically endless programming options. As such, it works well as a performative site where events, programs, and rituals occur and where the Jewish community is able to constantly represent and reproduce itself.

It is estimated that about 80 per cent of the Jews in Metro Detroit come to the JCC to participate in at least one program or event each year (Sheskin 2006). Certainly there are numerous social interactions that happen in the planning and attending of each event that contribute to a stronger sense of community. The importance of these should not be underestimated. Yet, these limited interactions seem to contribute more to the imagined or desired communal connectedness than the actual purposeful interaction and formation of bonds between individuals. It is at this level that the Kahn facility seems to fall short, and it is arguably due in large part to its architecture. Because Kahn was designed to be a massive programmatic facility, it works far less well as a meeting place where informal social contact can occur.

Even just approaching the building, it is easy to get a sense of the imposing scale. The parking lots are huge and set at a distance from the building. In part, this decision was made, and maintained even during the more recent renovations, due to security concerns. Nonetheless, one does not feel that one is approaching a welcoming community space. This is particularly true for seniors and people with disabilities for whom the distance is a significant barrier, especially in the ice and snow of the long Detroit winters. Upon arrival at the main entrance to the building, one enters the reception area and lobby, a dark, cavernous, double-story space that is usually empty and has the effect of making an individual feel quite small and alone. As the executive director, Mark Lit, stated, "[t]here could be a thousand people in the building, and you would never know it by looking at the lobby" (2008). While such spaces are commonly created to be awe-inspiring and make impressive symbolic statements about the power of the Jewish community, they can also be rather alienating.

The JCC leadership has tried to address the lobby problem. Not long ago, they moved and expanded the reception desk so that visitors could not pass without being greeted by the receptionist. They also tried shielding the lobby from view with portable screens and advertisements placed strategically behind the desk. Additionally, when possible, the Center tries to energize the space by using it for planned events and activities. As a large, visible, open space close to the front entrance, it is ideal for making the Center appear more active than it often really is.

Delving further into the building, it becomes clear that the wide winding corridors are both a problem and an asset when it comes to interaction and wayfinding throughout the building. As with many older buildings that have undergone numerous renovations over the years, circulation has become complex. Visitors can rarely find their way without assistance, and being lost reinforces the sense of being an outsider. Yet, in an odd way, this is also somewhat helpful. First, it forces newcomers to ask for directions, which stimulates a form of interaction. Secondly, it seems to give old-timers who do know their way around a sense of ownership of the building and belonging to the community.

The building as a whole is so large and underused that it often feels empty. There simply are not enough inhabitants during a typical day to make the building appear vibrant. The Center leadership has therefore tried several different strategic and behavioral tactics to stimulate interaction and render the building more welcoming and engaging. For example, the staff has been asked to follow certain rules while walking through the corridors of the building: if they come within ten feet of another person they are asked to make eye contact, and within three feet they should verbally greet or engage them (Lit 2008). Other solutions to the overwhelming scale have been to section off the building into smaller, more manageable pieces by programming and then to keep those limited areas more active. For instance, the upper floor was redesigned and then leased to the Frankel Jewish Academy day school. In addition to bringing in much needed revenue, this added a strong Jewish presence and more than 200 children to the building during school hours—the time when the JCC is most often empty.

There are, of course, some cases where the scale of the building can be seen as helpful in forming social relationships. Two seniors that I spoke to told me that they come to Kahn because of the wide range of opportunities all under one roof. They said that in the course of a day they can see their grandchildren, exercise, and attend Jewish or educational programming for retirees. For them, each point of activity and social interaction becomes a node in an expansive physical and social network. Moreover, the movement between these nodes is a way of physically engaging with and laying claim to the environment as their community space. But, overall, there are remarkably few intimate spaces in the building that would lend themselves to informal meetings and friendly interaction. The building lacks small public seating areas or spaces scaled to human (or bodily) proportions, where it would feel comfortable or appropriate to have a close conversation. For some time there was an upscale kosher restaurant in the building, but Center users found it to be too formal and expensive for just casual meals or coffee, and it closed due to lack of business. It was recently replaced with a pizza shop, which may better serve the interactive needs of the community. Not surprisingly, my informal surveys of people in the building suggested that most come for programming or to exercise and not to just "hang out." Several people also added that they bring friends or family members with them to programs because it can be difficult to meet or connect with new people. Interestingly, the most common place mentioned for meeting new people was the locker room, one of the few densely spaced, and undoubtedly less formal areas, where people were likely to find themselves in close proximity to others.

In many ways, when it comes to social interaction, JPM is the opposite of Kahn. The words that I most commonly heard used to describe JPM were *haimish*[2] and friendly, but a little run down. For some people the shabby condition makes it a less desirable symbol of an imagined community, but for others it seems more down-

---

2   Yiddish word defined as having qualities associated with a homelike atmosphere: simple, warm, relaxed, cozy, unpretentious, or the like (Haimish 2009).

to-earth and appropriate to its function—a comfortable site for sustained social interaction and for "real" community building. Although the building has a 1950s institutional quality, recent interior renovations have helped it to appear warmer and more welcoming. Visitors have only a short walk from the parking lot before they enter a bright, colorful lobby, which is central to the success of the Center. There is a hospitable front desk staffed by a receptionist who greets visitors, often by name, and directs them either to a small seating area or to other parts of the building. The desk staff also displays and sells candy and snacks, and helps with transactions of a video lending library, a gift shop, and a small "thrift shop." All of these functions give the lobby an active, engaging, and unpretentious quality that helps it to feel like the hub of the Center and, in many ways, of the community.

Both due to its smaller size and the fact that JPM is frequented by seniors and large Orthodox families with young children, many of whom have flexible daytime schedules, the building seems far more active and interactive than Kahn. Moreover, unlike the Kahn facility, the building seems much closer to human scale. It is easily navigated, with several places where one might stop to interact or chat with others. In addition to lobby seating and small available rooms such as the lounge, two places that stand out are the kosher café and the fitness center. The café has a relaxed, informal atmosphere making it an inviting place to have coffee or lunch before or after an event or workout. One is also much more likely to meet and talk to other people at JPM's fitness center than at Kahn's. While people are often frustrated at the small size of the fitness center, it is actually quite helpful when it comes to social interaction. To begin with, the limited size and close proximity of the equipment in the fitness center makes it difficult to avoid interaction with the people around one. Secondly, because it is smaller and less well-equipped than Kahn's or other gyms in the area, it is also more affordable. The moderate costs and welcoming community atmosphere have made it a good option for many Jews and non-Jews living in the vicinity. In this way, the fitness center is able to bring people of diverse backgrounds into contact with each other.

Jewish Identity

To some degree, all communities larger than a very small village are "imagined communities" based on a shared identity (see Anderson 2006). These communities rely on symbols, cultural products, and collective mythologies to sustain their communal identity. Buildings such as the JCC are cultural products that can play a dual function in an imagined community. On the one hand, they can become sites that foster interaction, helping imagined communal connections become realized. But, on the other hand, they can also be largely symbolic representations of cultural values and narratives that help to sustain imagined identities, in lieu of more constant social interaction.

The problem is that it is nearly impossible to create a non-denominational Jewish institution without entering the extensive modern debate on "who" or "what" is Jewish. Disagreements abound on such issues as family lineage, acceptance of Jewish law, religious practice, Zionism, and cultural markers, to name just a few. While these conflicts may seem trivial, an institutional stance will ultimately determine how the community will be represented and resources distributed. Therefore, it will also determine the proportion of the Jewish public that will feel comfortable and identified within a particular environment. In trying to market itself as a Jewish environment for the whole community, the leadership must largely try to either reconcile or sidestep these debates within its walls.

One must also keep in mind that the JCC is dependent on revenue from outside funding sources such as non-Jewish users and grants, many of which demand that the Center be open to any interested persons regardless of religious affiliation. Therefore, the JCC must try to find a balance between promoting and making Jewish culture available and obvious within the building environment, while still appearing to be a non-coercive institution. As such, the JCC seems to take three major approaches to fostering Jewish identity: 1) through the location of its buildings within the urban fabric, 2) by attempting to provide multiple Jewish options from which users may choose, and 3) by creating an appearance of Jewish unity by universalizing and normalizing American Jewish values and experience.

While neither Kahn nor JPM is architecturally distinctive on the exterior as a Jewish institution, both assert their Jewish identity on the interior and by their placement within the local geography. Both are located on Jewish campuses along with Jewish senior housing and other communal institutions. Kahn's campus is situated in the part of West Bloomfield referred to locally as Synagogue Row, due to the significant number of synagogues all within close proximity. Likewise, the JPM building is located in the heart of the restored Oak Park Jewish community. It is within walking distance of many Jewish homes, businesses, and communal institutions, with active Jewish cultural streetscapes.

Inside the two buildings, the JCC tries to make clear that one has entered a Jewish environment. Of course, the difficulty is in not overstating what is "Jewish." One way that the JCC tries to address and accommodate the wide variety of Jewish identities in the Detroit community is by providing numerous options from which people can choose. The most obvious choice is between the two buildings. Although amenities and convenience of location certainly play a large part, the decision-making process is also influenced by the reflection of values within the two buildings.

For example, there are a number of spaces throughout both buildings that are dedicated to "self-serve" Judaism. The library at the Kahn building is a particularly good example of this. Whereas the libraries in many synagogues and JCCs around the country are dark, outdated, and unused repositories, this one has been thoughtfully modernized with interesting architectural elements such as a curved wall design, an attractive entrance with large paned windows, bright new lighting, and comfortable reading areas. This library provides a quiet, relaxed atmosphere for individuals and families to spend time finding literature, music, or information that appeals to their sense of identity and religious observance.

The JPM building takes a similar approach (albeit on a smaller, cheaper scale) with its "Traditions Cart." This cart is located in the lobby next to the small seating area and is filled with flyers and information on Jewish holidays and traditions. Visitors to the Center can help themselves to reading material for immediate perusal or to take home.

Options for observance can also be found in areas such as the fitness centers. While this would seem to be a non-religious part of the building, any activities that focus on the body bring up Jewish issues of modesty and the separation of genders. If one looks closely at the glass walls of the aerobic studios at Kahn, they will notice that the back portion of the glass wall has been frosted. This is to allow religiously observant women to exercise with at least some degree of privacy. Subtle details such as these are an attempt at compromise between accommodating the Orthodox community and refusing to be overrun by their strict demands. But one person I spoke to at the Center mentioned that this was, in her opinion, an inadequate gesture that at best forced these women to the back of the room and at worst was simply insufficient to provide the required modesty. Some women mentioned that they were feeling as though the Center was forcing the observant community to use the JPM facility.

Unlike the Kahn building, most exercise rooms at JPM have solid walls and doors, as well as shades that can be pulled over the windows. There is also a freestanding wall in the lobby that blocks the glass windows overlooking the pool. These architectural details lend themselves more easily to a separate gender program schedule, and therefore more of this kind of programming happens at JPM rather than Kahn—much to the dismay of non-observant members who would prefer a less restricted schedule, especially during peak hours. While the two buildings do provide multiple options, their respective schedules and locations are inconvenient for many, and, perhaps more importantly, may create a division among identified Jews where more progressive Jews use Kahn and the more observant Jews use JPM.

As a third major strategy, the JCC has tried to downplay such religious divides and make the community seem more unified than it actually is. To do this, they take a two-pronged approach: they attempt to normalize the Jewish experience as one of many distinctive immigrant groups that has made significant contributions to local or national history, and they appeal to more "universal" Jewish values that are likely to be uncontested and have broad popular appeal. One of the primary ways that this is accomplished in the JCC environment is by framing Jewish activities as artistic and cultural contributions rather than religious behavior.

Since the early 1900s Jews have tried to normalize the Jewish experience and elevate the conception of Jews among themselves and in other outside groups by "fram[ing] the presentation of Jewish subjects in the cosmopolitan terms of art and civilization."[3] (Kirshenblatt-Gimblett 1996: 5). It seems that the JCC, even as it stands today, still follows this tradition. When I asked the executive director, Mark

---

3   Kirshenblatt-Gimblett explains that the history of Jewish ethnographic display was largely informed by the work of Cyrus Adler (1863–1940). Adler, though a religious

Lit, whether he thought that there was a dominant narrative that people read when they come to the Center, he answered that "I hope that you would see that there is a love of art and culture here" (2008).

Much of the Jewish content of the JCC environment comes in the form of museums or museum-like displays of art and Jewish contributions to Michigan's history, particularly in the Kahn building. This is likely because, as Steven Dubin explains in *Displays of Power: Memory and Amnesia in the American Museum*,

> [m]useums are important venues in which a society can define itself and present itself publicly. Museums *solidify* culture, endow it with tangibility, in a way few other things do. Unflattering, embarrassing or dissonant viewpoints are typically unwanted. (1999: 3, italics in the original)

In addition to the ubiquitous displays of art throughout the Center, there are at least four museum-type areas of the building: the Janice Charach Museum Gallery, which features rotating exhibits of art with Jewish themes or work by Jewish artists, the Michigan Jewish War Veterans Memorial (also at JPM), the Michigan Jewish Sports Hall of Fame, and Shalom Street, the interactive children's museum. Shalom Street is a noteworthy example because, in addition to being an ethnographic museum, it also includes the second characteristic appeal to more universal Jewish ethics and values. This museum is divided into four areas, each of which highlight communal and cultural aspects of Jewish life such as environmentalism, caring for one's neighbors, beautifying everyday life and rituals, and appreciating the sensory experience of being in a Jewish home. By shaping exhibits around cultural (even arguably atypical) values and practices, the JCC can avoid some of the more controversial and divisive religious values relating to ancestry, practice, and ritual, while encouraging broader Jewish identification and participation.

A similar approach has been taken in bringing Jewish content to other common areas of the building. For example, quotes from the Talmud about the essential indisputable values of good health and care for the body have been decoratively painted on the walls of the fitness center. There is also a graphic display, much like a stage set, of the city of Jerusalem, which has been hung from the second floor balcony above the lobby. It has the effect of making people standing in the lobby feel as though they are surrounded by the cityscape. Although there are certainly internal disputes within the Jewish community about Zionism, Israeli policy, and support for the peace process, the majority of American Jews consider their affinity with Israel to be a significant part of their Jewish identity and support its right to national sovereignty (Brit Tzedek v'Shalom 2010). The display highlights the relationship to Israel and Jerusalem as an important part of American Jewish identity without taking a strong political stance or engaging in a debate that might alienate potential members.

---

man who was sympathetic to the role of faith and religious practice, preferred Jews to be perceived as a highly cultured and civilized immigrant group.

Social Change

There are at least two ways to plan for social change: the first is to plan for what you hope and the second to plan for what you expect to see. Particularly for the JCC, deciding which approach to take seems to be a genuine dilemma. On the one hand, the explicit mission of the Center is to "ensure Jewish continuity." This would suggest efforts to attract and maintain the affiliation of the younger generation in Jewish institutions like the JCC. However, according to the Jewish Population Study, Detroit's Jewish population is aging. Planning for their future and the expected social change of the community would involve finding more ways to meet the needs of seniors, keeping them active, healthy, and involved. Consequently, the JCC has largely decided to divide and conquer. Consistent with the character and scale of the two campuses, the Kahn building seems to be trying to entice the younger generation, while the JPM building has continued to focus on the needs of its seniors.

With a hopeful eye towards the future, the Kahn building just completed another million-dollar renovation. This time the goal was to turn space previously used as Detroit's Holocaust Memorial Center (HMC) into the Beverly Prentis Wagner Teen Center. This seems to represent a major shift in the attitude, approach, and message of the JCC. Several years ago HMC moved to a separate facility a few miles away, leaving the JCC with a sizable empty area. In a building already underused and underprogrammed, the HMC's former location was not considered a desirable space. Actually, it somehow seemed appropriate for its previous function. Located below grade, it was cold and dark, with solid brick walls, no windows, and buzzing fluorescent lights overhead. The atmosphere was somber and reflective.

The Teen Center officially opened in March 2009, several months after my last visit. On my last tour of the space, in October, 2008, it had been in its final construction phase. While the space still felt a lot like a dark basement, it now did so in a "cool-teenage-hangout-sort-of-way." The space had been transformed and energized with bright colors, funky geometric shapes, and warmer, modern lighting. The former movie theater that once screened survivor testimonies has retained its function but now features newer technology, colorful seating, and shows pop culture films. The small sitting areas with televisions that showed documentary footage of the Nazi horrors are now game stations with flatscreen monitors, PlayStations, and Wii. Also added were a new snack bar and pool tables as well as new computers and quiet study carrels. The hope behind these changes is to render the space active and engaging by outfitting it with high end technology and design elements attractive to teenagers. Youths hereby would begin to feel a sense of ownership and connection through the Teen Center—and, by extension, to the JCC and the greater Jewish community.

The narrative of this space, for those who are aware of its history, is also quite powerful. For starters, one cannot ignore the message that a Holocaust museum documenting the plan to destroy the entirety of the Jewish people will now hopefully be reinhabited by a new generation of engaged Jewish teens. It also represents a major shift in the thinking of the Jewish community. Where the Holocaust was once cited as a major incentive for keeping Jews affiliated, that approach seems to hold less sway with the younger generation (see Dershowitz 1998). Today it seems that if active affiliation is the goal, Jewish communal institutions must find ways to stay relevant and keep people coming. Therefore, a state-of-the-art teen center is more likely to entice than a history lesson. It also clearly says in a visual and prominent way that the JCC prioritizes and invests in its younger members and sees them as the future.

Although research for the Strategic Plan of JPM showed that there were far more teens in the catchment area than previously thought, this demographic change has not been made a priority just yet. When I asked Leslee Magidson, Assistant Director to the Center and Manager of the JPM building, what changes had been made to the Center since my last visit, she excitedly told me about the new changing room added to the pool area (Magidson 2008). While a seemingly modest addition, particularly in comparison to the million-dollar teen center in the Kahn building, it does make an important statement about the priorities of the JPM building.

To begin, one must again consider the scale of the JPM building. At a packed 60,000 square feet and not much room to grow, choices for "found" space are not made lightly. But this choice is consistent with the goals of this building, which are largely to serve and engage the special needs of its population, including the elderly and disabled. The new changing room is intended to provide a dignified atmosphere for members who require the assistance of an aide or a family member, particularly when they are of different genders. Carved out of a storage closet alongside the pool deck, it is a plain, functional room that was designed to be fully handicapped accessible, with adequate space and privacy. There is nothing exceptional or even inviting about the space—no fancy finishes, lighting, or amenities. But from a practical standpoint, this changing room is already meeting the needs of several JCC members and will be an important way to keep the growing population of seniors actively using the pool and fitness facilities, even beyond the time that they begin to require additional assistance. While a seemingly small change, it will likely go a long way toward meeting the future needs of an aging community.

Although they differ somewhat in approach, the additions to both Kahn and JPM were made with an eye towards the future, and the intention of keeping as many of Detroit's Jews actively using the Center and engaged in the community for as long as possible. Both buildings have acknowledged a need to also plan for opposite demographics: Kahn needs to accommodate its seniors, many of whom are struggling with facility access, and JPM needs to plan for its growing population of teens. Yet, for the time being, while working with shrinking resources, Kahn's addition seems like a hopeful investment in a desired change, JPM's seems like a practical investment in an expected change.

Conclusion

Communities are created in a number of different ways: as a result of social interaction and group bonding, through desire and imaginings of community, and through active planning for future change. All of these can be achieved, at least in part, through manipulation of a social environment such as that of the JCC of Metro Detroit. When it comes to social interaction, the JPM building seems to be far more successful than the Kahn building, primarily because of its more manageable scale and opportunities for informal interaction. Kahn unfortunately falls into the trap of many large institutions. While the facility is certainly impressive, it tends to sacrifice "real" communal bonding in favor of a desired or imagined Jewish identity.

Still, for many, community is more about an idea or a feeling of connectedness to something bigger than themselves. Given the diversity of the Jewish community, fostering a sense of unity is no easy task. Here, Kahn has the advantage of size and resources, but both buildings have attempted to meet the challenge of creating a communal Jewish identity, largely by highlighting common denominators and framing them as universal and/or cultural values. Where this is simply not possible, Kahn and JPM try to offer a range of choices through location, architectural details, Jewish content, and scheduling. Additionally, the JCC leadership has tried to incorporate issues of interaction, community building, and identity in their plans for future change in the Detroit Jewish community. Seemingly unsure whether it is best to be hopeful or practical, they tried both. In a show of optimism, the Kahn building expended significant resources to reach out to the next generation of young Jews. JPM focused on the predicted needs of the growing elderly community, hoping to keep them active and involved at the Center for as long as possible. Through environmental considerations, both buildings have tried to stay relevant and meet the needs of their constituents for as far into the future as possible.

Works Cited

Anderson, B. 2006. *Imagined Communities*. New York: Verso.

Brit Tzedek v'Shalom: Jewish Alliance for Justice and Peace. 2010. *Resources— American Polls*. [Online]. Available at: http://btvshalom.org/resources/am_ polls.shtml [accessed: 9 June 2010].

Dershowitz, A. M. 1998. *The Vanishing American Jew*. New York: Touchstone.

Dubin, S. 1999. *Displays of Power: Memory and Amnesia in the American Museum*. New York: New York University Press.

Fortier, A.-M. 2006. Community, belonging and intimate ethnicity. *Modern Italy*, 11(1): 63–77.

Greenberg, H. 1975. *A Welcoming Word from the President*. West Bloomfield, MI: Jewish Community Center of Metropolitan Detroit.

Haimish, 2009, in *Webster's New World College Dictionary* [Online]. Available at: www.yourdictionary.com/haimish [accessed: 25 March 2009].

Jewish Community Center, 1973a. *News Release: New Center Plans Unveiled.* Detroit, September 14.

Jewish Community Center, 1973b. *Revised Proposal for the Construction of the Maple-Drake Building of the Jewish Community Center.* Detroit, May 1.

Jewish Community Center of Metropolitan Detroit, The Jimmy Prentis Morris Building, 2007. *Strategic Planning Process: "A Long-Range Vision."* Oak Park, MI.

Kirshenblatt-Gimblett, B. 1998. *Destination Culture: Tourism, Museums, and Heritage.* Berkeley: University of California Press.

Lit, M. 2008, interviewed by the author, 23 October.

Magidson, L. 2008, interviewed by the author, 23 October.

Peterson, R. A. 1979. Revitalizing the culture concept. *Annual Review of Sociology* 5, 137–66.

Sarfatti-Larson, M. 1997. Reading architecture in the Holocaust Memorial Museum: a method and empirical illustration, in *Sociology to Cultural Studies: New Perspectives*, edited by E. Long. Malden, MA: Blackwell, 62–91.

Sheskin, I. M. 2006. *The 2005 Detroit Jewish Population Study* [Online: Jewish Federation of Metropolitan Detroit]. Available at: http://www.jewishdatabank. org [accessed: 9 June 2010].

Wuthnow, R. 1994. *Producing the Sacred: An Essay on Public Religion.* Urbana, IL: University of Illinois Press.

Chapter 7

"Ambiguously Ethnic" in Sherman Alexie's Seattle: Re-Imagining Indigenous Identities in the Twenty-First Century

Ruxandra Rădulescu

Problematizing Postcolonial Seattle

In a world that has become increasingly polarized between global cities and the rest—according to the academic paradigm outlined by Anthony King (2003: 266)—attention is being drawn to the deployment of ethnicity as an added advantage in the race for "ethnic heritage tourism" (Gotham 2007: 125). As the advertising promises of a global capitalist metropolis are geared towards a reconstituted, class-oriented but quite flexible tourist market, the reinvention of tradition, of an exotic ethnicity and of a promising multicultural citizenship of the world appear appealing to global subjects always on the move.

Urban replanning in a world of global cities refocuses the interest in local and indigenous identities from being antithetical to (Western) social and scientific progress to being part of the cultural capital of global metropolises. Instead of being erased so as to make room for the advancement of Western modernity, indigenous cultures are now rebranded as significant factors in the urban identity of the place. Place-stories, to use Coll Thrush's coinage in *Native Seattle* (2007) are being told to emphasize heterogeneity in a heterotopian world precisely in the face of impending delocalization and homogenization.

Multiethnicity and world-famous indigeneity serve as major indicators of urban exceptionalism as embedded in the self-presentation of Seattle, one of the nodal centers of the Pacific Northwest region. As historians, novelists, and politicians agree (Thrush 2007: 3–16), Seattle is a city marked by a multiplicity of signifiers of ethnicity, mostly pointing to the past, to a spectral Seattle of indigenous phantasms, and yet also by a gesturing toward the future, a possible horizon of reclaimed space and place. Being entirely negativistic about the present of urban Native Americans or naively optimistic about forthcoming recognition would set up another easy dichotomy, which would prevent a more nuanced look at current developments. Instead, this chapter proposes a more segmented view upon postcolonial urban cultural politics in the fiction of a Native American Seattleite, Spokane/Coeur d'Alene writer Sherman Alexie, drawing on insights from Postcolonial Literary Studies and global city politics.

In this light, it could be argued that marketing Seattle relies on two core political dimensions: Seattle as a postcolonial urban space and as a global metropolis of capitalism. Its growing importance, from an outpost on the urban frontier in the hinterlands of the Pacific Northwest to a regional urban power to a global business center, in the logic of financial progression, has been highlighted in several studies pointing to its place in United States and world politics, including the grassroots anti-globalization struggles that marked the World Trade Organization (WTO) forum in 1999. Seattle's position of prominence in the postfordist economy of software services as well as its steady growth of influence as a major Pacific Northwest hub of world commerce have all been acknowledged, and yet Seattle has also been both advertised and actively reclaimed as an aboriginal space.

Thus, centuries-long struggles to recognize the existence and participation of Native populations in the construction of the city are both rendered visible and coopted as marketing devices in the political impact of the city. As former mayor Wes Uhlman admitted, being able to provide visitors with a tour of the Skid Row area in downtown Seattle that included a view of urban Indian residents of the area was a touristic bonus of indigenous authenticity (Thrush 2007: 177). Talking about Seattle therefore necessarily involves attending to the dispossession and displacement of Native populations, migration, diaspora, and neocolonialism, on the one hand, and radical urban activism, survival, resistance, and endurance, on the other.

However, addressing Seattle's ongoing colonial history is no easy task. Two methodological problems occur: urban scholarship has only begun to approach the political and cultural status of postcolonial cities, and in case a postcolonial status is acknowledged, it is all too readily assumed to be only a matter of excavating the past, without regard for the carving of an indigenous presence in the urban space. As Nicholas Blomley correctly complains, "with a few important exceptions, the recognition of the city as a postcolonial space has been limited" (2004: 108). Urban Studies, one of the fastest growing clusters of disciplines with relevance to a very shifting postmodern world, has yet to come to terms with a palimpsestic urban text replete with imbricate alternative histories, Homi Bhabha's ambivalent postcolonial temporalities, and sectional cultural differences. Similarly, anthropological scholarship of American Indian populations has yet to seriously address the social histories of urban Indians beyond the urban/reservation divide. Each contributing discipline, from Urban History to Race Studies to Postcolonial Studies, can nonetheless participate in a larger paradigm of intersectionality so as to facilitate interrogating a multiply determined Native presence in a Westernized urbanized space that has been inhabited uninterruptedly by indigenous peoples for thousands of years.

Authenticating the Urban, Commodifying the Indigenous

To investigate the dimensions of an indigenous urban space and indigenous urban cultural politics, this chapter will focus on a number of aspects regarding the commodification of indigenous identities in the United States, as well as on moments of disruption and redrawing of boundaries that may describe the US not as postethnic, but interethnic as well as postcolonial, as identifiable in the work of Spokane/Coeur d'Alene author Sherman Alexie. The fantasy of playing Indian, which Philip Deloria analyzes so thoroughly in *Playing Indian* (1999), has defined the very core of identity-forming processes in the new republic. In his study, Deloria explains the dialectic of imagining both "interior" Indians—the free-spirited and wise noble savages—and "exterior" Indians—the savage and primitive members of the vanishing race. While claiming that an interior Indianness could help the revolutionaries legitimize their rebellion by playing the card of indigeneity against the United Kingdom, the ideology of an exterior type of repulsive and rejected Indianness helped justify the westward expansion and ensuing land theft on the part of the United States government.

Employed to either purpose, playing Indian has involved the fabrication of a homogenous indigenous identity—or, as Gerald Vizenor calls it, "simulations of Indian identity"—threatening to once more displace and replace indigenous populations with a suitable and marketable racial profile. The cultural politics behind the stereotyping, as benevolent as it might be, of American Indian identity may be seen in the concluding lines of Sherman Alexie's poem *How to Write the Great American Indian Novel*. The poem takes to task suitable and tragically comic stereotypes of the romantic, stoic-looking Indian, such as the following:

> All of the Indians must have tragic features: tragic noses, eyes, and arms.
> Their hands and fingers must be tragic when they reach for tragic food.
> The hero must be a half-breed, half white and half Indian, preferably
> from a horse culture. He should often weep alone. That is mandatory. (1996a: 94)

The poem ends more dramatically, however, with an image that both continues and contradicts well-known lines from the alleged 1854 speech of Chief Seattle. Alexie states: "In the Great American Indian novel, when it is finally written / all of the white people will be Indians and all of the Indians will be ghosts" (1996a: 95), where the spectral character of the American Indian may very well represent its own substitution by white simulations of Indianness. In the alleged speech of Chief Seattle, the image of the Indian as ghost is not necessarily connoted as a position of victimhood:

> And when the last Red Man shall have perished, and the memory of my tribe shall
> have become a myth among the White Men, these shores will swarm with the
> invisible dead of my tribe, and when your children's children think themselves

alone in the field, the store, the shop, upon the highway, or in the silence of the
pathless woods, they will not be alone. (Seattle 2000: 729)

The image of a haunting Indian presence, beyond conveying a sense of defeatism,
is also an instance of a very Derridean spectral presence, of the irreducible and
enduring difference, of an uncanny and uncontrollable supplement that disturbs
the politics of the center.

However, a deconstructive reading of the Indian uncanny would require
yet another presentation. Instead, various accounts of the branding of Seattle
demonstrate the commodification of indigenous Seattle, a city marked by an
iconography of totem poles, statues of Indian patrons, and pavements in the
pattern of cedar-bark baskets (Thrush 2007: 3). Established in the second half
of the nineteenth century, Seattle is a palimpsestic city that weaves together
multiple pasts of a diversity of indigenous tribes and a number of governmental
policies marking the westward expansion. In need of a defining identity, it resorts
to playing Indian again when, in 1889, the city welcomes its trademark totem
pole. Characteristically, though, for a colonial nation, the totem pole had been
stolen from a Tlingit village in Alaska, where a party of Seattleites enacted a brutal
moment of cultural imperialism. One of the participants in the expedition gave the
following account of the negotiations with the natives:

> The Indians were all away fishing, except for one who stayed in his house and
> looked scared to death. We picked out the best looking totem pole ... I took a
> couple of sailors ashore and we chopped it down—just like you'd chop down
> a tree. It was too big to roll down the beach, so we sawed it in two. (quoted in
> Wilma 2000)

The head of the expedition party explained the success of the mission:

> I was, at the time, the Acting President of the Chamber of Commerce and the
> excursion was largely under my direction. ... Arriving at the point designated,
> we found an abandoned Indian village with probably 100 or more totem poles,
> from which 17 had been taken a year previously by the Harriman Expedition
> and distributed to different colleges throughout the East. I should have stated
> that there were two decrepit Indians which we finally succeeded in interviewing,
> who made no objection to our taking the pole to Seattle. In fact, the Indians were
> as pleased at our taking it as were the people of Seattle to receive this outstanding
> example of workmanship of the Northern Indians. (quoted in Wilma 2000)

Imagining thus a felicitous communicative act that empowered the American
migrants and further symbolically alienated the resident natives, the act of stealing
a prospective urban crest—Seattle's mark of distinction—highlights precisely
the refounding of a community through appropriation and reauthentication.
Dismembering the totem pole not only robs the indigenous artisan community

of its coherence and symbolic foundation but also denies agency to its creators as well as its native recipients in Seattle. The heritage it means to tout is one doubly marked by impersonation: the members of the expedition party replay the role of indigenous craftsmen by redoing the totem pole, and the city of Seattle gains a distinctive identity by displacing its own Duwamish and Suquamish native heritage and replacing it with a beautified one from an entirely different cultural area.

Discussing the politics of global tourism, Kevin Fox Gotham (2007: 126) observes that one development in contemporary ethnic heritage tourism consists in attracting immigrants that may be able to help establish trendy new ethnic districts. Branding Seattle as a city with a *rewritten* indigenous core carries out the process of displacing the dispossessed. Building on Nicholas Blomley's thesis, the argument that dispossession is not complete without both physical and symbolic displacement helps clarify the necessity for an eclectic mix of indigenous markers of identity too disparately and haphazardly put together to convey a narrative of enduring and consistent indigenous presence.

Literature in a Reclaimed City

In a post-9/11 world of heightened US-American nationalist sensitivity, the challenge of multiculturalism and the question of postethnicity have drawn responses ranging from a celebration of a healing type of modern-day ethical cosmopolitanism to fierce critiques of imperialism as globalization. The negotiation of ethnic positionalities continues to be played out on the contested terrain of urban cohabitation, where some communities are redefined by intercultural alliances, while others are locked in confrontation.

Signaling his trickster discourse of in-betweenness, the phrase "ambiguously ethnic," used by Spokane writer Sherman Alexie in *Ten Little Indians* (2004), is a controversial (self-deconstructing) label that challenges received or desired notions of ethnic authenticity, particularly in an urban geography. Set in the city of Seattle, marked by a history of late colonization and a strong indigenous presence, the stories of *Ten Little Indians* convey a complex understanding of indigeneity in the twenty-first century. This essay will argue that the multiple cultural allegiances of Alexie's characters are not a sign of naïve cosmopolitan idealism, but follow a logic of contestation and "violence of hybridity" (Patell 1997), as foregrounded in Alexie's novel *Indian Killer* (1996b). As such, the narrative landscape/urban ethnoscape Alexie's texts construct is one fraught with challenges, dramatic failures, and parodying plays on popular culture stereotypes in a "postindian" redefinition of community.

It is not only in fiction but also in film that Sherman Alexie subverts expectations of commodified indigeneity. In his 2002 movie *The Business of Fancydancing* the opening scene shows a young, successful, gay poet, Seymour Polatkin, of the Spokane tribe, giving a reading of his poem *How to Write the Great American*

Indian Novel in the window of a bookstore by a sign that says "American Indian Heritage Month." During the reading of the poem Seymour is shown kissing the bust of Chief Seattle on the lips. The image of the bookstore window Indian addressing no other audience but the implicit spectators of the movie underscores precisely the objectification of indigeneity in a space of painful emptiness—no one is physically attending the reading in the bookstore—that is, at the same time, a space of cinematic presence. The scene moreover both challenges gender-based notions of Indianness, with Seymour Polatkin kissing chief Seattle in an emphatically erotic manner, and raises questions regarding the musealization of indigenous people.

However, literary representations of Seattle have seen an arguably marked change in Sherman Alexie's work. In *Indian Killer* Seattle is the scene of multiple murders, signed by an unknown murderer, which prompted critic Cyrus Patell (1997), building on previous doubts about living as a Native American in a white city as raised in Leslie Marmon Silko's novel *Ceremony*, to talk about the "violence of hybridity" in Alexie's fiction. The main character of *Indian Killer*, an Indian man named John Smith, had been raised by a white adoptive family, but had grown to develop a paranoid schizoid behavior as a consequence of being trapped in two seemingly incompatible worlds. Nonetheless, Alexie's later approaches to urban Indianness evince a more comprehensive view on composite, hybrid global identities.

In Alexie's collection of short stories *Ten Little Indians* urban Indians appear less violent and filled with rage. Instead, the stories present readers with a complex range of Indian characters that range from a graduate student in love with poetry to a basketball player to a series of homeless Indians. It is precisely the peregrinations of homeless Indians reinscribing the city with a renewed network of cooperation and intercultural alliance that can be read as indicative of a restorative sense of identity through survival and continuance.

The question of the homeless Indians in Seattle is no easy question at all. Pushed away by the waves of settlers, Indians now total less than one per cent of the population of Seattle. Many of them are destitute and homeless and have been living in the downtown Skid Road area, Seattle's "Indian territory," which happens to be located around Pioneer Square, the historic district of the city. Historian Coll Thrush describes it as "an urban neighborhood with its own traditions, institutions, and ways of operating" (2007: 174). During the urban renewal phase, the city council put forth a project of restoration of its historic center, which involved shutting down hotels and driving Indian residents out of the reputed Skid Road. In 1972 Mexican Indian writer J.A. Correa thus surmised with regard to the musealization traditions in the historic district:

> Pawn shops
> and broken people
> who drink and sing
> and beg for wine
> they provide amusement
> for the tourists

who believe in
historical sites
and the little kids
from school
are taken by
devoted bored teachers
to see the heart of their
grandparents' city. (qtd. in Thrush 2007: 178).

Sherman Alexie's self-reflexive short story *Imagining the Reservation*, published in *The Lone Ranger and Tonto Fistfight in Heaven* in 1993 and bitterly commenting on the predominantly unilateral infusion of mainstream popular culture into reservation life, describes the situation of urban Indians as essentially defined by mainstream media: "Imagine Crazy Horse invented the atom bomb in 1876 and detonated it over Washington, D.C. Would the urban Indians still be sprawled around the one-room apartment in the cable television reservation?" (1993: 149). Circumscribing his representation of the Spokane Indian reservation within the parameters allowed by the stereotypes of the culture industry, Alexie asks: "What do you believe in? Does every Indian depend on Hollywood for a twentieth-century vision?" (1993: 151) and proposes imagination with an ironic twist as a way to escape the "terminal creeds" (Vizenor 1994: 120) of Hollywood's romanticization of the so-called vanishing race.

Despite Alexie's dry humor, his urban Indians are nonetheless caught in a destructive cycle of hypotheses and failed vision quests. The image of the enclosed Indians who perform their identity in a reservation of the mind as constructed by the media reflects another act of symbolic displacement. While, technically, there are no urban Indian reservations with borders that either protect or displace or both, as in a traditional reservation (itself only a political construct), media constructions of Indianness confine the indigenous subjects to a static life of futile fantasies of revenge. In this sense, Alexie's fundamental insight in *The Lone Ranger and Tonto Fistfight in Heaven* is that the urban environment, despite its promise of democratization and freedom of choice, is one more site of seclusion and pauperization. The contrasting images of the sprawling Indians and of their one-room residences are parallel to the feeling of enclosure many tribes experienced during forced relocation to reservations.

What is also fundamental about Alexie's representation of Seattle is that it does not point to any obvious sites of exclusion in the manner of the all too familiar confrontational positioning of ghettoes versus suburbs. Unlike Pomo writer Greg Sarris, Alexie does not discuss the insider/outsider dialectic as being a product of spatial relations within a definite environment. Alexie's urban reservation is everywhere, from bars to Skid Row in the historic district to Pike Place and university campuses. As such, it engages more faithfully Henri Lefebvre's discussion of space as a product of social relations, where meandering through the nooks and crannies of a city founded in direct cooperation with the resident

indigenous populations determines the fluid contours of the city, which will
become more evident in *Ten Little Indians*:

> The urban is, therefore, pure form; a place of encounter, assembly, simultaneity.
> This form has no specific content, but is a center of attraction and life. It is an
> abstraction, but unlike a metaphysical entity, the urban is a concrete abstraction,
> associated with practice ... What does the city create? Nothing. It centralizes
> creation. Any yet it creates everything. Nothing exists without exchange,
> without union, without proximity, that is, without relationships. The city creates
> a situation, where different things occur one after another and do not exist
> separately but according to their differences. The urban, which is indifferent
> to each difference it contains, ... itself unites them. In this sense, the city
> constructs, identifies, and sets free the essence of social relationships. (Lefebvre
> 2003: 117–19)

The one-room urban reservation is also a comment on the relocation process
that left many Native Americans confused, deprived of family connections, and
crowding in inadequate housing. As historian Donald Fixico explains, urban life
has led to the formation of new stereotypes of destitute, homeless, alcoholic Indians
(2006: 31), which aligned with long-standing negative images of the maladjusted
native Other. Feelings of inadequacy soon developed, thus adding new layers to
the already internalized justification of displacement as a necessary process of
Westernization.

Alexie's constant denaturalization of the highly traditional concept of "tribe"
and tribal connections, now rewritten as flexible worldwide communities and
temporary rhizomatic associations, disrupts the essentialist narrative of self-
contained indigenous communities, whose members share a specific understanding
of land and definite tribal descent. The new type of rootedness in multinational
patterns of survival that some of his middle upper-class characters experience is
the rooted cosmopolitanism that defines the ethics of identity in the global world,
in keeping with Anthony Appiah's understanding of the interconnectedness of
traveling cultural subjects who need not feel brutally uprooted in order to be
part of a dynamic and forever expanding world. The loss of traditional roots,
while disconcerting and leading to tragedies for individuals whose existence is
endangered by the contemporary wave of instability, is not always already a sign
of decadence and of the impossibility of healing.

In Alexie's much-acclaimed short story *What You Pawn I Will Redeem* the
main character Jackson Jackson, a homeless Indian that is painfully aware of his
continuing nomadism, redefines the history of Spokane and Seattle by inscribing
it with the enduring presence of his people:

> I'm a Spokane boy, an Interior Salish, and my people have lived within a hundred-
> mile radius of Spokane, Washington, for at least ten thousand years. I grew up
> in Spokane, moved to Seattle twenty-three years ago for college, flunked out

within two semesters, married two or three times, fathered two or three kids, and then went crazy. (Alexie 2004: 169)

Personal breakdown and tribal displacement combine in the story to provide a background not only of expected failure, but also of idiosyncratic success, which is both marginal and triumphantly collaborative.

Jackson Jackson, living on the edges of society even though in a community of other homeless Indians with relatives in far better conditions, encounters and vanquishes imperialistic commodification of Indianness in his own way. Spending time with his drinking companions in Pike Place Market in the historic district, Jackson sees his grandmother's dance regalia in a pawnshop window and tries to redeem them by raising the sum of money the shop owner requests. As he goes about earning and spending money with inexhaustible largesse, Jackson cannot raise the money within the initial deadline of 24 hours and yet is ultimately gifted back his grandmother's regalia by the merchant.

In his quest to redeem a piece of commodified heritage for sale in downtown Seattle, Jackson questions the codes of cultural conversion of the dancing regalia from a ceremonial attire to a marketable item, once more pointing out the depletion of the indigenous cultural patrimony. However, while Jackson's predicament is a comment on the economic condition of the invisibility of the working-class indigenous members of the urban community, here included through the figures of three Aleut workers who simply vanish in the course of three pages, his quest for cultural survival and revival is supported by an entire multicultural network of friends and acquaintances that range from a white police officer to a Korean female sales attendant to the multitribal community in Big Heart's bar. To the extent to which Jackson fails to control the flow of money he needs to reach his target sum, his friends, his acquaintances, and other people he has never met before coalesce to form an informal support system.

In a combination of fairytale motifs and Pacific Northwest indigenous potlatch rituals, the story suggests cooperation and communal striving toward a worthy goal. To obtain the prize (his very own heritage and thus a significant portion of his own self), the hero must pass through a number of tests and overcome obstacles with the help of his multicultural friends, which lends itself to a narratological analysis looking to underline elements of folktale morphology. Moreover, even as the hero fails to produce what is expected of him on a literal level, his incapacity to hold on to money translates precisely as his fundamental adherence to a revived potlatch tradition of the Pacific Northwest. Spending his money on his friends and random strangers, Jackson defies the logic of accumulation and enacts the potlatch pattern of sharing as much a part of his heritage as his grandmother's regalia. Finally, Jackson's impulse to dance with his grandmother's powwow regalia in a road intersection highlights the potential for celebrating cultural restoration in disjunctive places turned suddenly conjunctive, in an otherwise alienating environment that is now made to listen to the rhythm of a recovered native tradition:

I took my grandmother's regalia and walked outside. I knew that solitary yellow bead was part of me. I knew I was that yellow bead in part. Outside, I wrapped myself in my grandmother's regalia and breathed her in. I stepped off the sidewalk and into the intersection. Pedestrians stopped. Cars stopped. The city stopped. They all watched me dance with my grandmother. I was my grandmother, dancing. (Alexie 2004: 194)

While Sherman Alexie does not indulge in an unlimited celebration of the assumedly liberating multicultural practices of the global metropolis, his texts are structured around negotiations of the historical presence and construction of indigeneity in the cosmopolitan regime of multinational capitalism. Even if no single discourse dominates the scene of literary productions, in line with postmodern heterotopias, the strength of Alexie's recent contributions highlights the potential for exploring the emergence of a new Native American presence in the global flows of the city.

Conclusion

Postcolonial approaches to contemporary US-American indigenous literatures have generally met with skepticism, given that ongoing colonialism is still considered a defining factor in the relations between the United States government and Native American communities as "domestic dependent nations." However, postcolonial theory is not limited to the legal definitions of a chronologically ordered process of colonization. It can serve as backdrop for Diaspora Studies examining the ways in which Native American urban residents of the first and second generation— that is, the generations that were relocated in the 1950s and their urban-born descendents—relate to one another, to members of their communities back on the reservation, and to other Native American urban inhabitants in order to form a multiplicity of connections that acknowledge the trauma of exile as well as the strength of urban activism. This is where contemporary urban-centered literature can come into play by (re)claiming the space that indigenous people have already carved out for themselves in the city as well as challenging the canon of US-American literature that is often little welcoming toward Native American authors who do not write about reservation life.

Stories of urban planning and replanning can no longer avoid the history of urban expansion by means of excluding as well as strategically including its original residents. This seems particularly appropriate in the context of United States urban enterprises, originally represented as crucial to the westward advancement of civilization. On the one hand, postcolonial theory in the global city can disrupt the very process of Westernizing history and destabilize the codes of representation that place Native Americans at the outskirts of recent history. It can cut through layers of repression and national mythology laying bare the structures of public invisibility that shatter the foundation of a harmonious global conglomerate, as it is done, for example, in Sherman Alexie's disturbing revenge novel *Indian Killer*. On the other hand, Native American literature by and about indigenous urban residents typically exposes the masquerade of placing cities under the symbolic leadership of Native figures such as Chief Seattle as a way to underline the marketable ethnicity of exotic Indians. Stolen tribal regalia and artifacts that decorate the city are markers of an imperialist fantasy of ultimate urban control over the wilderness of the frontier, of a mascot-like presence of historic and historical Indians. Yet, the possibility of personal and collective interethnic redemption such as in *What You Pawn I Will Redeem* may still subversively come out of the rejection of commodification, through a city-wide effort to change the flow of money into a (somewhat idealized, but still no less inspiring) flow of multiethnic aspirations and dreams.

Works Cited

Alexie, S. 1993. *The Lone Ranger and Tonto Fistfight in Heaven*. New York: Harper Perennial.

Alexie, S. 1996a. How to write the great American Indian novel, in *The Summer of Black Widows*. New York: Hanging Loose Press, 94–95.

Alexie, S. 1996b. *Indian Killer*. New York: Warner Books.

Alexie, S. 2004. *Ten Little Indians*. New York: Grove Press.

Blomley, N. 2004. *Urban Land and the Politics of Property*. New York: Routledge.

Business of Fancydancing, The (dir. Sherman Alexie, 2002).

Chadwick, A. 1999. Blood and memory. *American Literature*, 71(1), 93–116.

Deloria, P. 1999. *Playing Indian*. New Haven: Yale University Press.

Fixico, D. 2006. *Daily Life of Native Americans in the Twentieth Century*. Westport: Greenwood Press.

Gotham, K. F. 2007. Ethnic heritage tourism and global-local connections in New Orleans, in *Tourism, Ethnic Diversity, and the City*, edited by J. Rath. New York: Routledge, 125–39.

King, A. D. 2003. Postcolonialism, representation and the city, in *A Companion to the City*, edited by G. Bridge and S. Watson. Oxford: Blackwell, 261–9.

Lefebvre, H. 2003. *The Urban Revolution*, translated by R. Bonnono. Minneapolis: University of Minnesota Press.

Patell, C. 1997. The violence of hybridity in Silko and Alexie. *Journal of American Studies of Turkey*, 6(1), 3–9.

Seattle. [1854] 2000. The Indians' night promises to be dark, in *Encyclopedia of Minorities in American Politics*, vol. 2: *Hispanic Americans and Native Americans*, edited by J. D. Schultz et al. Phoenix: Oryx Press, 728–9.

Thrush, C. 2007. *Native Seattle. Histories from the Crossing-Over Place*. Seattle: University of Washington Press.

Vizenor, G. 1994. *Manifest Manners: Postindian Warriors of Survivance*. Hanover, NH: Wesleyan University Press.

Wilma, D. 2000. Stolen totem pole unveiled in Seattle's Pioneer Square on October 18, 1899, in *HistoryLink*. [Online: Washington State University]. Available at: http://www.historylink.org/index.cfm?DisplayPage=output.cfm&file_id=2076 [accessed: 7 June 2010].

Chapter 8

Against the "Erasure of Memory" in Los Angeles City Planning: Strategies of Re-Ethnicizing LA in Digital Fiction

Jens Martin Gurr and Martin Butler

An area is not a slum because the people living there are poor and black. A community becomes a slum when the space is represented for everyone (including those who live there) by those who don't live there.

Marback 1998: 82

The twin beasts that erased much of downtown [Los Angeles]—racist neglect and ruthless planning—leave only a faint echo in cinema, because generally one will distract the other, or because cinema, by its very apparatus, resembles the tourist imaginary.

Klein 2008: 249

We have attempted to develop a non-metaphorical notion of culture as text. Concepts such as the archive as a full-text database and the search request, which are vital to this view, suggest the thought of actually converting our theory of textuality into a method of analysis based on data-processing.

Baßler 2005: 294

Los Angeles, of course, has long been a center of attention for urbanists as well as for scholars of urban planning and of cultural representations of the city. It has been the subject of innumerable studies, the locale for countless novels, documentary films, and particularly of countless feature films.[1] However, one of the most impressive attempts to render the complexities of twentieth-century Los Angeles, which does justice to these complexities by means of a highly self-conscious form of presenting a wealth of material, is Norman M. Klein's multimedia docufiction *Bleeding Through: Layers of Los Angeles 1920–1986*. It combines a 37–page

---

1   From among the innumerable studies see for instance Davis 2006, Fulton 2001, Klein 2008, Murphet 2001, Ofner and Siefen 2008, Scott and Soja 1997, Soja 1996a, 1996b, and 2000.

novella by Klein with a multimedia documentary DVD[2] based on Klein's research on twentieth-century Los Angeles.

Bleeding Through is loosely based on the fictitious story of "Molly," who moved to LA in 1920 when she was 22, and whose life and times the narrator of the novella attempts to chronicle. The question whether or not she killed her second husband (or had him killed) at some point in 1959 serves as a fake narrative hook to launch the reader and user of the DVD on a quest through layers of twentieth-century Los Angeles, especially through the area around Molly's fictitious home in Angelino Heights. The work thus centers on a few surrounding neighborhoods close to downtown Los Angeles such as Bunker Hill, Boyle Heights, Chavez Ravine, and Chinatown, which were all demolished in questionable urban development projects in the second half of the twentieth century. The impetus of the project is archaeological in nature in that it attempts to re-present these erased parts of the city—not only in the sense of torn-down buildings and entire neighborhoods, but in the more complete sense of the social fabric and the lives and realities of the people who lived there, entire *Lebenswelten* that have also disappeared in such processes of urban renovation. This area, the setting of countless screen murders in Hollywood films, has also been the site of rampant real estate speculation and the concomitant racist eviction policies. *Bleeding Through* thus explores the connection between genre-specific urban imaginaries of Los Angeles, racist urban policies, greed, and the concomitant ruthless urban planning.

In this chapter, we draw on basic assumptions of urban cultural studies, ecocriticism, and cultural ecology in order to stress the active and "regenerative" role of urban cultural forms of expression. After having thus explored what we would like to call the seismographic and the catalytic momentum of cultural forms of expression in urban environments, we will establish some of the contexts for an analysis of Klein's multimedia documentary by discussing some of the key findings of Klein's 1997 monograph *The History of Forgetting: Los Angeles and the Erasure of Memory*, on which *Bleeding Through* is based to a considerable extent. By outlining the relationship between the novella and the DVD and by highlighting some of the features and design principles of the interactive DVD, we will then lay the foundations for a discussion of how *Bleeding Through* re-presents twentieth-century Los Angeles. We will show how *Bleeding Through* takes us on a revisionist tour of twentieth-century LA history and points out the extent to which film and fiction (the innumerable *films noirs*, detective films, and thrillers set in LA) have shaped perceptions of the city. It thus provides a subversive view of hegemonic strategies of urban planning and ethnic segregation and undercuts conventional representations of multiethnic LA. We will then more explicitly comment on some of the implications of the hypertextual, multimedia format of the documentary, arguing that *Bleeding Through* makes full use of the opportunities afforded by the digital medium of what one might call "interactive documentary database fiction

---

2   References to the novella, where this source is not clear, will be abbreviated as *BT*, references to the DVD will be given by tier and chapter.

in hypertext format." In our conclusion, we propose that Klein's work nicely lends itself to illustrating the notion of culture as an "archive" (Baßler 2005) and the perceived convergence of cultural theory and technological developments in hypertext as a medium of cultural memory. However, we will also argue that it is precisely the solitary nature of the experience of navigating these complexities which deprives the work of some of its assumed subversive potential.

Urban Culture as Seismograph and Catalyst: The Regenerative and Commemorative Function of Urban Cultural Forms of Expression[3] and the Archival Function of Urban Hypertexts

In order to contextualize our arguments about the function of Klein's hypertext in countering the erasure of memory and in remembering multiethnic twentieth-century Los Angeles, we would like to begin with a brief exploration of some of the basic assumptions about the overall significance of urban culture that have become central to the field of Urban Cultural Studies. On the basis of these assumptions, and with reference to major arguments from the field of ecocriticism, we will then sketch a model of urban culture, conceptualizing it as a "force-field" that bears diagnostic as well as regenerative potential in urban systems and is thus, as our analysis of *Bleeding Through* aims to show, particularly useful for a "thick description" of urban cultural practices and forms of expression.

One function of urban culture that was long underestimated but has recently received wide publicity is its economic importance as a major economic sector and as a location factor in attracting economic and cultural "elites." One might here think of Richard Florida's widely debated but partly simplistic and problematic theses on the "creative class."[4] In addition to potentially overstating the contribution of specific forms of culture to an attractive economic milieu—theses which have led a number of cities strategically to target "creative segments" of the population in their urban development strategies—, the concomitant instrumentalization of art and artists has also met with significant resistance with artists refusing to be commodified as mere location factors conducive to the "bohemian index" of a city.[5]

The significance of urban culture to the city as a whole, however, not only emerges from its instrumental value and its measurable and immediate economic

---

3   Some of the ideas set out in this section have been inspired by the research carried out in the University of Duisburg-Essen's Main Research Area "Urban Systems," in which we are both centrally involved. In interdisciplinary cooperation across virtually all departments of the university, over 70 scientists and scholars in numerous multidisciplinary projects here engage with key issues in urban systems.

4   See for instance Florida 2004 and 2005. For the use of such factors in city marketing cf. for instance Gold and Ward 1994, Kearns and Philo 1993.

5   See for instance the much-publicized protest of artists in Hamburg against such endeavours, "Kunst als Protest" (2009).

relevance. Even more significant is its social and socio-psychological function as a form of critically negotiating social, political, and economic problems and particularly as a medium for critical reflection on processes of urban development and change as well as on the limitations and restrictions set by highly technologized and functionalized urban settings. Moreover, urban cultural practices serve as ways of expressing and articulating individual and collective identities, which, particularly in urban agglomerations, where cultures and ethnicities constantly mix and mingle, seems to be of vital importance.

Given the intricate interplay between urban culture and its environment, we believe that it is most fruitful to conceive of urban culture as a quasi-ecological system, which, as a dynamic and cybernetic entity, develops according to its own logic and rules.[6] We claim that such an understanding of urban culture, which takes up central ideas of some of the most recent strands of ecocriticism, 1) allows us systematically to conceptualize the dynamic interplay between urban cultures and their environments, (also) because it allows for the integration of a number of different disciplinary approaches, and 2) thus provides us with a theoretical framework that may serve as the starting point to "literalize" the metaphor of "cultural ecology" and to underline the role of Urban Cultural Studies in processes fostering sustainable urban development.

To begin with, it would seem that, given the strong interest in ecocritical approaches and in urban studies in recent years or even decades, an application of ecocritical concepts to the study of the metropolis lies close at hand. However, while some forays into this domain have been made (Bennett 2001, Bennett and Teague 1999), most studies in ecocriticism—both classics in the field and more recent work—have remarkably little to say about *urban* cultures. We therefore propose to heed Michael Bennett's still pertinent warning that "ecocriticism will continue to be a relatively pale and undertheorized field unless and until it more freely ventures into urban environments" (2001: 304).[7]

One approach from this field that particularly lends itself to a conceptualization of urban culture as an ecological system is the model of literature as cultural ecology outlined by Hubert Zapf, which, though it explicitly focuses on literary texts, is particularly useful for the analysis and description of the dialectical and quasi-ecological relationship between forms of cultural expression and their specific contexts. We maintain that, by way of a few terminological and conceptual modifications, it is thus also transferable to the realm of urban culture, which may well be conceived of as a dynamic ecological system subject to constant change, too.

---

6   We are aware that, in the comparison between an ecological system and culture, the term "ecological" most often is only employed and understood metaphorically. However, it is one of our aims in this contribution to show that and how "cultural ecology" can also be taken literally.

7   Buell (2005: 23) has correctly, we believe, remarked on due attention to urban concerns as a key feature distinguishing what he calls "second-wave ecocriticism" from the more narrow first-wave ecocriticism primarily concerned with nature writing.

In his approach, Zapf outlines a functional theory of literary texts that is based on the assumption that the system of literature in many respects resembles an ecological system (2001: 90–92). Enumerating a number of striking analogies between the two, particularly highlighting their dynamic and complex nature,[8] he concludes that "the specific procedures of literature bear some interesting similarities to … ecological principles … Indeed, they appear to a significant extent as the transformation into language and symbolic action of some of those characteristic principles" (2001: 90). Zapf also points out that the similarity between an ecological system and literature is predominantly due to the specific aesthetic strategies employed by the literary imagination, when he claims that

> literature is an ecological force within culture not only or not even primarily because of its content, but because of the specific way in which it has evolved as a unique form of textuality that, in its aesthetic transformation of cultural experience, employs procedures in many ways analogous to ecological principles, restoring complexity, vitality and creativity to the discourses of its cultural world by symbolically reconnecting them with elemental forces and processes of life—in non-human nature, in the collective and individual psyche, in the human body. (2001: 93)

Following from this, the symbolic system of literature turns into a socially and culturally productive agent

> and, by its aestheticising transgression of immediate referentiality, becomes an ecological force-field within culture, a subversive yet regenerative semiotic energy which, though emerging from and responding to a given sociohistorical situation, still gains relative independence as it unfolds the counter-discursive potential of the imagination in the symbolic act of reconnecting abstract cultural realities to concrete life processes. (2001: 88)

Starting from his notion of literature as an "ecological force-field within culture," Zapf then argues that "this cultural-ecological function of literature can be described as a combination of three main purposes" (93), which he further specifies as those of a "cultural-critical metadiscourse," an "imaginative counter-discourse," and a "reintegrative inter-discourse." Both these three functions and his metaphor of a "force-field" bear a particular potential for a systematic analysis of urban culture. This potential lies in the possibility of transferring the idea of "literature as an ecological force-field within culture," which lies at the heart of Zapf's approach, to another level of metaphorical abstraction and of applying it to an urban context. Consequently, and in accordance with Zapf's idea, we may well conceive of "urban culture as an ecological force-field within

---

8   For a detailed elaboration on these analogies, see Zapf 2001, 88–90. See also Zapf 2002, 2005: 60–62, and Zapf 2008.

urban systems." Urban cultural practices reflect and comment upon forms and functions of architectural and infrastructural designs and thus may be said to work as a cultural-critical metadiscourse. In processes of constant transgression and subversion, they create alternative spaces and thus function as an imaginary (and, at times, very concrete) counter-discourse. Subversive as such practices may be, they frequently only exist on, at, or in (and thus due to) a specific architectural or infrastructural given; that is, they, by definition, reintegrate the ideologically peripheral with manifestations of hegemonic power in the very moment of their being produced, installed, or performed. Thus, considering the processual and performative nature of urban cultural practices, they may well be characterized both by a "seismographic momentum" in that they react to, or "track," urban transformations in a very sensitive way, and a "catalytic momentum" in that they actively interpret, make sense of, foster, and even instigate technological, infrastructural, and cultural changes and challenges.

Urban cultural forms of expression thus understood can conceivably occur in an enormous range of different shapes. It therefore seems that a further heuristic category may be fruitful in the systematic exploration of "urban tactics" (de Certeau 1984) of appropriating cityscapes. We here propose to speak of the "mediacy" of the urban cultural practice employed by individual or collective urban players to appropriate and negotiate the spatial dimension of their very specific urban environment. This relational category allows a further specification of the forms and functions of urban cultural expression, as it provides a continuum between more direct and more indirect forms of negotiating urban space.

Among the more direct "tactics" of appropriating and redefining urban spaces are, for example, forms of performance art that explore the architectural and technological constraints and possibilities of urban environments, but also specifically urban forms of sports such as skateboarding, BMX cycling, or parcouring. A further form of immediate engagement with pre-structured urban environments may be seen in the subversive appropriation of walls, roofs, or streets as "canvases" for graffiti and other forms of street art, which, more often than not, do not only redefine urban spaces by changing the surface structure through coloring and iconic as well as non-referential forms of expression, but may also contribute to establishing feelings of a shared (ethnic) identity among a particular collective, for example, by deliberately undermining established versions of colonial history, thus re-writing the past and subverting hegemonic ideologies.

In addition to these more direct forms of coming to terms with urban environments, there are more indirect forms of "dealing with" the city in fictional and non-fictional texts—in the broadest sense—ranging from literature to city guides, via the daily news on television or the blockbuster about 9/11 to the multimedia hypertext on urban developments that forms the subject of this chapter. Such medial representations do not only articulate particular perspectives on the metropolis and/or give a voice to their inhabitants, but may also render dystopian or utopian urban scenarios, "possible spaces," so to speak, which make us aware of (hypothetical) consequences of processes of urbanization.

What this short enumeration of some examples once more helps to illustrate is that both the mediate and immediate tactics of appropriating or negotiating the metropolis must not be exclusively conceived of as (critical or non-critical) reactions to urban spaces, as *seismographs*, so to speak, tracking urban developments and changes in a very sensitive manner. On the contrary, assuming that urban cultural forms of expression do indeed constitute a "force-field" within the larger infrastructural, technological, and architectural framework of the metropolis, we believe that these tactics, as a kind of *catalyst*, also contribute to shaping our view of the city and thus, being socially and culturally productive, potentially have a significant impact on our understanding and perception of the environment most of us live in.[9] Moreover, it is not only our perception of the city that is altered by cultural forms of expression which use or represent urban spaces. It is the actual development of the city itself that is closely tied to medial representations and appropriations of the metropolis, as Manfred Faßler (2006) points out, going so far as to state that "urban developments are historically inseparable from media evolutions" (Faßler 2006: 21, our translation).[10]

In the course of this chapter, we would like to use Klein's *Bleeding Through* as a case study to discuss ways in which hypertextual database fiction as a highly mediated form of urban cultural expression can be seen to diagnose key urban challenges as well as actively to shape perceptions of urban environments and thus to be socially productive as well. Finally, we aim to show how certain forms of urban cultural commentary reintegrate the repressed into the cultural memory of the city.

Bleeding Through, "the Erasure of Memory," and the Multimedial Reconstruction of Twentieth-Century Multiethnic LA

In Klein's 1997 book *The History of Forgetting: Los Angeles and the Erasure of Memory*, on which *Bleeding Through* is based to some extent, he names as one of his key themes "the uneven decay of an Anglo identity in Los Angeles, how the instability of white hegemonic culture leads to bizarre over-reactions in

---

9   For the production of city images and the effects of such fictional and documentary images of the city, of self-projected images of different types of cities ("world cities" such as London, New York, and Tokyo; "wannabee world cities such as Paris, Chicago and Toronto"), or of projected images such as that of being "clean and green" or having successfully managed the transition from the industrial to the service economy ("Look! No more factories" [Short 1996: 431]) upon the perception of cities see Short 1996: 414–62. See also the other contributions in this volume.

10   For the close correspondence between views of the city and developments in literature and the arts see several of the contributions in Smuda 1992. Lobsien and Smuda (1992), in particular, elaborate on the connections between modernism and urbanity. For the way in which fictional images of a city—especially filmic images—come to physically shape the real city, see Klein 1998, 2008.

urban planning, in policing, and how these are mystified in mass culture" (Klein 2008: 17). Referring to what is probably the most emblematic and most drastic urban redevelopment project in twentieth-century Los Angeles, Klein states that "Bunker Hill [became] the emblem of urban blight in Los Angeles, the primary target for redevelopment downtown from the late twenties on" (2008: 52). The Bunker Hill Renewal Project, begun in the 1950s and not scheduled to end before 2015, brought the virtually total razing of a neighborhood, the flattening of the hill and the building of the high-rise buildings now popularly regarded as constituting "Downtown LA." Similarly, Klein comments on the razing of the "old Chinatown, old Mexican Sonora, … the old Victorian slum district, and other *barrios* west of downtown [that] were leveled, virtually without a trace" (2005: 97). Commenting on an eerie commonality of all these twentieth-century urban renewal projects in LA, Klein writes:

> [E]xcept for Chinatown, every neighborhood erased by urban planning in and around downtown was Mexican, or was perceived that way (generally, they were mixed, often no more than 30% Mexican). … While East LA may *today* seem the singular capital of Mexican-American life in the city, the mental map was different in the forties. The heartland of Mexican-American Los Angeles was identified as sprawling west, directly past downtown, from north to south. Bunker Hill was identified as "Mexican" by 1940, like Sonoratown just north of it … and particularly Chavez Ravine. (Klein 2008: 132–3)

In this context, he also refers to the "policy of shutting out downtown to non-whites […] since the 1920s" (2008: 132).

In *History of Forgetting* as well as in a number of essays, Klein further shows how the urban imaginary of Los Angeles has been shaped by images of the city in film, from noir to *Bladerunner* and beyond, creating "places that never existed but are remembered anyway" (Klein 2001: 452), even arguing that the ideology of noir and neo-noir films, these "delusional journeys into panic and conservative white flight" also help "sell gated communities and "friendly" surveillance systems." (Klein 1998: 89).

Bleeding Through juxtaposes two formats: the "traditional" narrative of Klein's novella with a number of additional texts (mostly by Klein's collaborators on the project) on the one hand, and a multimedia DVD on the other. Klein's constantly self-reflexive 37-page novella "Bleeding Through" has a highly self-conscious first-person narrator who tells the story of Molly and—just as centrally—his own attempts to reconstruct it:

> I couldn't trust any of her stories. Not that her facts were wrong. Or that she didn't make an effort. [But] she'd fog out dozens of key facts. Whenever I noticed, she would blow me off, smiling, and say, "So I lose a few years." … But there were seven memories in the years from 1920 to 1986 that were luminously detailed. (Klein 2001: 10)

It is these seven memories of key stages of her life around which the novella and the DVD are structured. The underlying story of Molly and her life in LA between 1920 and 1986 thus serves to explore levels, layers, and developments in the city happening in these 66 years. The "preface," readable on the DVD above a vintage photograph of downtown LA with City Hall still by far the tallest building, makes clear the central principle of the "cinematic novel archive" of *Bleeding Through* (Klein 2001: 453) and already highlight its major concerns:

> An elderly woman living near downtown has lost the ability to distinguish day from night. Rumors suggest that decades ago, she had her second husband murdered. When asked, she indicates, quite cheerfully, that she has decided to forget all that: "I lose a few years."
> Three miles around where she is standing, more people have been "murdered" in famous crime films than anywhere else in the world. Imaginary murders clog the roof gutters. They hide beneath coats of paint. But in fact, the neighborhoods have seen something quite different than movie murders; a constant adjustment to Latinos, Japanese, Filipinos, Jews, Evangelicals, Chinese. What's more, in the sixties, hundreds of buildings were bulldozed. And yet, pockets remain almost unchanged since 1940. (DVD "Preface")

Set largely within the three-mile radius near downtown LA in which Molly spent most of her life, the documentary deals with neighborhoods such as Boyle Heights, Bunker Hill, Chavez Ravine, Chinatown, and Echo Park, the disappearance of which was chronicled in *History of Forgetting*. As the narrator explains here, this area was the site of the most drastic urban renewal projects in the country continuing over decades: "Hundreds of buildings gone: that could just as easily have been caused by carpet bombing, or a volcano erupting in the central business district" (*BT* 12). In the middle of these radically erased and unrecognizably rebuilt areas, six streets of Angelino Heights have remained largely unchanged since 1925. However, the area around Angelino Heights is also the center of a filmic universe: "Inside those three miles, under the skyline dropped by mistake into downtown ten years ago [in the 1980s] more people have been murdered in classic Hollywood crime films than anywhere else on earth" (*BT* 12)—the novella refers to "290 murder films ... shot no more than five minutes from Molly's house" (*BT* 31).[11]

The first "tier" of the documentary, "The Phantom of a Novel: Seven Moments," is structured around the seven key moments of Molly's life in LA between 1920 and 1986. Historical photographs of people and places in the neighborhoods surrounding Bunker Hill dominate these seven chapters. Thumbnails of the photographs are arranged in random sequence and can be selected by the user;

---

11   The narrator of the novella later states that "[s]ince the Seventies, murders have been relocated a few blocks west, because gunfire looks more ironic underneath the LA skyline at night, seen best from the hills in Temple-Beaudry" (*BT* 37).

alternatively, the user can go through the photo archive by enlarging each photo to almost the size of the screen and then moving on either to the photograph on the left or on the right. Making full use of the technical possibilities, the sequence of photographs is not fixed but rather randomly brought up from the archive. Additionally, with each phase of Molly's life, there is a short narrative comment by Norman Klein in a window in the corner the reader may open and close at will. The narrator of the novella describes this first tier as "a visual, interactive radio program … a kind of modern novel on a screen with hundreds of photos and Norman as narrator. You might say they are also a docu-fictional movie" (*BT* 43).

Tier Two, "The Writer's Back story," which the narrator of the novella describes as "more like a contextualization," is largely made up of newspaper clippings and establishes the context of other people and places more loosely connected to Molly's story. It collects newspaper clippings covering events and developments that occurred during Molly's life, with references to the prohibition and illegal distilleries, the ban on interracial marriages in the state of California in a 1932 newspaper clipping, the controversial reception of a 1941 anti-Semitic speech by Charles Lindbergh, the deportation of Japanese Americans during World War II, illicit gambling, or the McCarthy era with its Red Scare and the building of air-raid shelters—frequently interspersed with innumerable sensationalist clippings reporting murders in Los Angeles. Additionally, explanatory captions beneath newspaper clippings and photographs contextualize developments, with comments, for instance, on the ambivalent views of Chinatown in the 1920s as both "an exotic place in the popular imagination" and a location "considered as an eyesore, as more brown and black races converged at the Plaza" (DVD 2:1).

Tier Three, "Excavation: Digging behind the story and its locale," is described in the novella as "the aporia of media itself" (*BT* 43). In five sections, it offers a wealth of further material, here arranged thematically rather than chronologically. A section entitled "People Molly Never Met But Would Make Good Characters in Her Story" features randomly arranged interviews with twelve actual residents (including Norman Klein) of these neighborhoods who comment on their experiences within the social and ethnic developments in twentieth-century LA, the Zoot Suit Riots, fear of violent police officers, ethnic festivities, anti-Communist witch-hunts during the McCarthy era, the 1947 murders of Elizabeth Short—the "Black Dahlia"—and of Bugsy Siegel, or the treatment of Japanese Americans during World War II. Largely consisting of film and video sequences, it is a "vast 'ironic index' of what Molly left out, forgot, couldn't see. It samples from the back-story that gets lost when the movie or novel is made legible" (*BT* 43). It is also described as

> a meta-text (not a deconstruction). It is the structure of what cannot be found, what Molly decided to forget, what Molly never noticed, what passed before her but was lost to us. It is proof that no novel or film (documentary or fiction) can capture the fullness of how a city forgets, except by its erasures. (*BT* 38–9)

As the narrator of the novella comments, "each tier, then, comments on a specific medium that tries to make the city intelligible as it erases, collectively forgets, survives from day to day" (*BT* 43). The documentary database includes hundreds of images, frequently pairing an old photograph and a recent one taken from exactly the same angle; some of these are made to blend into one another in fascinating overlay montage. Drawing on various archives (the sources are scrupulously documented on the DVD), the documentary also makes use of maps, newspaper clippings, drawings and sketches, historical film clips, and (for copyright reasons) films snippets recognizably re-enacting key scenes of famous LA films merely by repeating the camera movements in basically empty streets in the original locations, but without actors,[12] numerous interviews with long-term residents, sometimes elaborate captions, as well as narrative commentary by Norman Klein. Klein's video commentary frequently gives clues as to the story behind the disappearance of Molly's second husband Walt, which adds a playful dimension of detective game to the navigation experience, because, such is the underlying fiction, the point of navigating *Bleeding Through* in the first place is to act as a detective on the hunt for such clues. However, as the narrator of the novella comments, "[t]he journey through the evidence is more exciting than the crime itself. We want to see everything that is erased to make the story legible" (*BT* 37).

Thus, neither the novella nor the DVD are to be regarded as a higher-level commentary on one another; they are mutually complementary: Just as the DVD can be seen as a vast exploration of the themes outlined in the novella, the narrative frequently comments on the contents of the DVD: "Next day, I went into a newspaper morgue, looking for articles on Walt's disappearance. Instead, I found fifty ways to kill a man between 1959 and 1961 (along with five suicides). I've scanned all the articles into a database for you" (*BT* 24). In the novella, the fictional story of Walt's disappearance is constantly related to current developments chronicled on the DVD, tying the wealth of documentary material back to the underlying quest narrative: "Among police photos, I find what should be Walt's body. ... Then I discover that on the same day, the downtown editor cancelled photos about racist crimes, particularly the railroading of blacks and Latinos" (*BT* 25).

Molly is herself a new-comer and an outsider when she arrives in the city in 1920, as Klein the narrator tells us: "In 1920, Molly is a twenty-something girl from a Jewish home in the Midwest, who arrives in Los Angeles to find her husband to be" (DVD 1:1; cf. also *BT* 13). As Delphine Bénézet points out, "[t]hrough Molly, Klein articulates a gendered and minority-oriented revision of the city's history" (2009: 69). With Molly as its protagonist, *Bleeding Through*

---

12   In addition to these re-shot sequences—frequently iconic scenes with the downtown towers looming in the background—of many murder films set in LA, this section also features maps of the city with key locations used in these films such as *Falling Down, Heat, Training Day, Chinatown, The Last Boy Scout, T-Men, Omega Man, Dead Or Alive,* or *To Live and Die in LA*

shifts attention from hegemonic white males and directs it to the role of minorities in LA's complex history.

From the very beginning of Klein's insert narrative on the DVD, Molly's neighborhood is characterized as a multiethnic one, when Klein refers to a family of "Latino's renting downstairs" (DVD Preface). Much of the material centers on transformations in twentieth-century multiethnic LA, whether in references to "Brooklyn Avenue with its famous mix of Jews and Mexican, Japanese and other 'swart' young men" (*BT* 15; cf. also 40) to "restrictions against the black community on Central Avenue, especially when by 1924 membership of the Klan reached its highest number ever" (*BT* 30), to the tearing down of Chinatown for Union Station (built in 1939), to the history of mixed Japanese and Mexican neighborhoods, with a Japanese American family man running a Mexican grocery store (DVD 3:1), the 1943 Zoot Suit riots, the Watts rebellion, or the turning of Little Tokyo into "Bronzeville," when African Americans and Mexicans moved into the area while the Japanese Americans were held in deportation camps away from the West Coast (*BT* 22).

The changes in twentieth-century Los Angeles are rendered in a fascinating if oblique way in the multiplicity of double photographs that morph into one another; as in a pair of photos taken on the corner of Spring and Main Street in the 1920s and today, in which a shop sign "D.W. Wong Co. Chinese Herbs" disappears and a billboard advertising Green River Bourbon morphs into a billboard advertising a $ 7,000,000 lottery draw in Spanish (DVD 1:2). In another of these morphs, juxtaposing 1941 Main Street with a contemporary image, "Fond's Pants Shop" on 655 Main Street (with "Ben's Barber Shop" and "Adams Radios & Appliances" next to it) turns into "Dongyang Machine Co." (DVD 1:3).

A central theme, however, are the more drastic changes imposed by radical urban development projects in areas such as Bunker Hill. The section "Collective Dissolve: Bunker Hill," with film sequences from Kent McKenzie's 1956 documentary *Bunker Hill* and *The Exiles* (1961), maps, and photographs from the 1890s to the 1960s attempts to recreate Bunker Hill before the massive demolition program that cleared the area for what is now regarded as "downtown" LA. A long sequence from McKenzie's *Bunker Hill* refers to the Community Redevelopment Agency's major redevelopment plan to relocate 8,000 residents of the neighborhood, to demolish all buildings, and to sell the land and have modern office and apartment complexes built (DVD 1:6). This chapter of the DVD also displays images from 1959 and 1960 showing the large-scale demolition of Bunker Hill. A sequence from Gene Petersen's 1949 film *... And Ten Thousand More* [housing units] also refers to the problem of "slums" in LA and the need for urban development. This sequence is captioned "The Myths of Urban Blight."[13] Similarly, the photograph of a model "Redevelopment Study for Bunker Hill, March 22, 1960" is captioned "Cooking Statistics to Justify Tearing Down Bunker

---

13   On the discourse of crisis and the frequently disastrous consequences of large-scale restructuring plans in LA cf. also Soja 1996b.

Hill" (DVD 1:6). Indeed, statistics on the housing situation and living conditions in Bunker Hill appear systematically to have been distorted in order to win public support for the demolition of this predominantly Mexican neighborhood. In the caption underneath a sequence from McKenzie's *The Exiles*, the fact that "this was a brown and black identified downtown center" is explicitly identified as "one of the reasons it was torn down" (DVD 3:3).

In the interview section, Japanese American Bill Shishima comments on his pre-war childhood in the Mexican community:

> We were welcomed in the Mexican community, so my Dad had a Mexican grocery store there by Olvera Street. I have a feeling that my Mom was more fluent in Spanish than in English because of necessity. Most of my neighbors were Mexicans or Hispanics, so that all I had was Mexican friends. (DVD 3:1)

In a later section, he reports on his experience of having to leave Los Angeles in May 1942 as an 11-year-old to be interned away from the coast with his family. He further comments on how:

> [d]uring the war, Little Tokyo was empty and vacant and so it became Bronzeville. Many people from the South came here to Los Angeles to work in the war industries ... Many of them just ended up here in First Street or Little Tokyo, and it was predominantly black Americans so they lived here in Little Tokyo and changed the name to Bronzeville. (DVD 3:1)[14]

Other residents similarly comment on racial segregation in LA. Thus, retired African-American fireman Arnett Hartsfield reports coming to Los Angeles in 1929, "when we couldn't even cross Washington Boulevard on Central Avenue [because of segregation]" (DVD 3:1). Similarly, Esther Raucher recalls her experience of first coming to downtown as a white child and of staring at African Americans: "As a child ... I don't think I'd seen a black person ... That's how segregated the city was that you would never see a black person" (DVD 3:1). Tying such developments to the underlying story of Molly, a clip from Jeremy Lezin's 1975 documentary *A Sense of Community* with references to illegal immigrants working in LA is captioned "With each year, Molly felt the massive immigration from Latin America change the rules in her world" (DVD 1:7).

All in all, in keeping with *The History of Forgetting*, on which it is based in substance, *Bleeding Through* thus shows how twentieth-century Los Angeles, in the process of becoming increasingly multiethnic demographically, continued to erase the visible traces of this diversity in favor of a de-ethnicized all-American

---

14   These experiences are echoed in a clip from Claude Bache's 1957 film on the treatment of Japanese Americans during World War II and their deportation to relocation camps away from the coast as well as on Japanese American volunteers in the US armed services during the war (DVD 2:5).

look and feel modeled on the needs of a largely white elite and enforced by representing ethnic LA along the lines of the paranoid and implicitly racist aesthetic of innumerable *noir* murder films.

Implications of the Form: Hypertext Docufiction as Radicalized Historiographic Metafiction and Political Commentary

The experience of navigating *Bleeding Through* is a fundamentally contradictory one: On the one hand, by making sophisticated use of the technological possibilities of the multimedia documentary, the fast-paced, multi-dimensional, overpowering, non-hierarchical, multi-faceted documentary recreates the urban experience of twentieth- or even twenty-first-century LA; on the other hand, there is a nostalgic quality to the experience, which partly arises from the use of vintage photographs, film clips, and newspaper clippings that appear to work against the grain of the high-tech mode of presentation—in keeping with Klein's views expressed in *The History of Forgetting: LA and the Erasure of Memory* on the constant self-reinvention of the city and the concomitant memoricide of previous layers of its history. However, while these aesthetic and experiential implications of the form are worth noting, even more momentous are its implicit politics, which elegantly complement the more explicit political commentary also packaged into *Bleeding Through*.

In contrast to even the most advanced filmic documentaries, which still inescapably rely on the linearity of film, *Bleeding Through*, makes full use of the digital medium to break linearity. Thus, while documentaries, which are originally meant for collective viewing, induce forms of collective medial experience, the effect of *Bleeding Through* specifically relies on a highly individual experience. The constant need to "do" something in the process of navigating the multiple layers—all clips are very short; hardly anything happens without being triggered by the user, who is essentially assigned the role of a detective in search of the truth—thus not only foregrounds the mediality, narrativity, and construction of the material, it also activates the viewer. In keeping with the promise of the medium,[15] the non-linear presentation of the material thus precludes closure, stimulates the discovery of knowledge rather than imposing it, and thus fosters learning without being explicitly didactic.

If, as Kathleen Burnett has argued, "[h]ypertext is rhizomorphic in all its characteristics" (Burnett 1993: 28)—and Klein's work makes full use of the hypertext medium—*Bleeding Through* may be characterized as fully rhizomatic, with all the non-totalizing and anti-hegemonic implications Deleuze and Guattari

---

15   For analyses of the technological and literary implications of the digital form and their repercussions in Literary Studies from the early 1990s classics to more recent accounts see Aarseth 1997, Bolter 2001, Burnett 1993, Ensslin 2007, Gaggi 1997, Hayles 2008, Landow 1997 and 2006, McGann 2001, Simanowski, Schäfer and Gendolla 2010, Sloane 2000.

famously ascribe to rhizomatic discourses.[16] Thus, the multimedial, multivocal, multi-perspectival, interactive, non-sequential, and highly self-reflexive experience of navigating *Bleeding Through* brings out "traits that are usually obscured by the enforced linearity of paper printing" (Burnett 1993: 3) and, like hypertext generally, serves to undercut, liquefy, and question established and hegemonic representations with their frequently unquestioned dichotomies and *hierarchies violentes* (*sensu* Derrida).

Database fictions, in flaunting the arbitrariness of such choices and enabling users to choose differently next time (but never exactly to retrace their steps), are potentially subversive purely in their form in that they expose as a construction and fabulation what narrative traditionally represents as a given. By making each journey through the material necessarily a different one—and by thus presenting what is merely material for a story as subject to change and human intervention— these narratives also contribute to the activation and mobilization of the user in ways that even the most advanced self-reflexive fiction—which is still subject to the unchangeable linearity of print—cannot achieve (cf. also Kinder 2003: 54).

True to the "democratic form" of hypertext digital media, *Bleeding Through* by means of its form thus already serves to deconstruct hegemonic constructions of history, as it constantly draws attention to the medial, discursive, constructed nature of such conceptions. As a user, one is never allowed to forget that this is a revisionist, anti-hegemonic, at times polemical re-construction of a repressed, alternative Los Angeles.

Conclusion: *Bleeding Through*, Archival Database Fiction, Urban Cultural Expression, and the Problem of Solitary Subversion

By juxtaposing the visions and grand designs of city planners with the voices of former inhabitants of razed neighborhoods, *Bleeding Through* implicitly takes up the perspective "from below," considering the city not so much as consisting of roads and buildings but of people, who—in the sense of de Certeau (1984)— tactically make use of the hegemonically imposed infrastructure of the city in the daily pursuit of their lives. Thus, by inviting readers to critically review processes of ruthless urban planning, Klein's archival database fiction not only functions as a "critical meta-discourse," but—by working against the "memoricide" implicit in the dominant discourse of urban planning—also works as an "imaginative counter-discourse." Finally, by focusing on what is normally repressed in hegemonic discourses of the city, it also fulfils the function of a "reintegrative inter-discourse" in Zapf's sense (2001). This memorial potential of Klein's medial configuration

---

16   As proposed by Deleuze and Guattari (2004: 7–13), rhizomes is characterized by the principles of "connectivity," "heterogeneity," "multiplicity," "a signifying rupture," "cartography", and "decalcomania." For a discussion of the rhizomatic nature of hypertext along the lines of these characteristics, cf. Burnett 1993.

of the city, we believe, particularly stems from its hypertextual structure. In order to conceptualize this archival and memorial function of literary and cultural production, we propose to draw on Moritz Baßler's notion of the archive, which he defines as follows:

> We will use the term archive to designate ... the sum of all texts of a culture available for an analysis. In the archive, these texts are accessible without being hierarchized. The archive is a corpus of texts. Within this corpus, passages equivalent to each other can be marked with a search request, as it were. These passages form an intertextual *structure of equivalence.* (Baßler 2005: 196, our translation)

Following George P. Landow's notion of a "convergence of contemporary critical theory and technology" (Landow 1997), Baßler then suggests that this theory of textuality, of intertextuality, and of text-context relations might quite literally be translated into a methodology of contextualizing cultural analysis based on data-processing technology (Baßler 2005: 294). Thus, he refers to the cultural archive in the sense of such a totality of available texts as a "full-text database" (2005: 293, our translation) accessible by means of search requests and organized in the manner of a hypertext. Indeed, Baßler's entire terminology and methodology suggest a view of the cultural archive as the collection of heterogeneous and not necessarily contemporary texts in a synchronic, non-hierarchically ordered hypertextual database. This "archive," it is easy to see, will fulfill an important function in maintaining and shaping cultural memory. This notion, one might argue, is literalized in *Bleeding Through*.

Engaging with various forms of appropriating the city—from the most direct version proceeding by means of bulldozers to fictionalized and medialized forms in filmic urban imaginaries—*Bleeding Through* is itself a form of appropriating the city, if a mediatized rather than an immediate one. As a documentary meta-appropriation of the city, it is a unique form of cultural expression in an urban context, in which it takes up a diagnostic as well as catalytic function, seeking to raise awareness and potentially to influence policy-making and urban planning by celebrating the ethnic and cultural diversity of Los Angeles in the face of ruthless urban redevelopment based on racism and greed. However, while *Bleeding Through* thus helps to counter the memoricide induced by urban planning in LA, the question remains to what extent a cultural product that so centrally relies on the individual, solitary user for its experiential form of negotiating central urban issues and which thus inherently forgoes any chance of fostering a sense of community can ever be truly subversive. Moreover, in order once more to highlight the centrality of agency and voice, one might ask who speaks, and on whose behalf.

Works Cited

Aarseth, E. J. 1997. *Cybertext: Perspectives on Ergodic Literature.* Baltimore and London: Johns Hopkins University Press.

Baßler, M. 2005. *Die kulturpoetische Funktion und das Archiv: Eine literaturwissenschaftliche Text-Kontext-Theorie.* Tübingen: Francke.

Bennett, M. 2001. From wide open spaces to metropolitan places: the urban challenge to ecocriticism. *ISLE: Interdisciplinary Studies in Literature and Environment,* 8(1), 31–52.

Bennett, M. and Teague, D. 1999. *The Nature of Cities: Ecocriticism and Urban Environments.* Tucson: University of Arizona Press.

Bénézet, D. 2009. Recombinant poetics, urban *flânerie,* and experimentation in the database narrative. *Bleeding Through: Layers of Los Angeles 1920– 1986. Convergence: The International Journal of Research into New Media Technologies,* 15(1), 55–74.

Bolter, J.D. 2001. *Writing Space: Computers, Hypertext, and the Remediation of Print.* 2nd Edition. Mahwah: Lawrence Erlbaum Associates.

Buell, L. 2005. *The Future of Environmental Criticism: Environmental Crisis and Literary Imagination.* Malden, MA: Blackwell.

Burnett, K. 1993. Toward a theory of hypertextual design. *Postmodern Culture* 3(2) [Online]. Available at: http://muse.jhu.edu/login?uri=/journals/postmodern _culture/v003%20/3.2burnett.html *[accessed: July 01, 2010]*

Davis, M. 2006. *City of Quartz: Excavating the Future in Los Angeles.* Brooklyn, NY: Verso.

de Certeau, M. 1984. *The Practice of Everday Life,* translated by S. Rendall. Berkeley: University of California Press.

Deleuze, G. and Guattari, F. 2004. Introduction: Rhizome, in *A Thousand Plateaus,* translated by B. Massumi. London and New York: Continuum. 2nd Edition, 3–28.

Ensslin, A. 2007. *Canonising Hypertext: Explorations and Constructions.* London: Continuum.

Faßler, M. 2006. Vorwort: Umbrüche des Städtischen, in *Die Zukunft des Städtischen: Urban Fictions,* edited by M. Faßler and C. Terkowski. Munich: Fink, 9–35.

Florida, R. 2004. *The Rise of the Creative Class: And How It's Transforming Work, Leisure, Community and Everyday Life.* New York: Basic Books.

Florida, R. 2005. *Cities and the Creative Class.* New York: Routledge.

Fulton, W. 2001. *The Reluctant Metropolis: The Politics of Urban Growth in Los Angeles.* Baltimore: Johns Hopkins University Press.

Gaggi, S. 1997. *From Text to Hypertext: Decentering the Subject in Fiction, Film, the Visual Arts, and Electronic Media.* Philadelphia: University of Pennsylvania Press.

Gold, J.R. and Ward, S.V., (eds.) 1994. *Place Promotion: The Use of Publicity and Marketing to Sell Towns and Regions.* Chichester and New York: Wiley.

Hayles, N.K. 2008. *Electronic Literature: New Horizons for the Literary*. Notre Dame: University of Notre Dame Press.

Kearns, G., and Philo, C., (eds.) 1993. *Selling Places: The City as Cultural Capital, Past and Present*. Oxford: Pergamon.

Kinder, M. 2003. Bleeding through: database fiction, in *Bleeding Through: Layers of Los Angeles 1920-1986* [DVD and Book], with additional texts by R. Comella, M. Kinder, A. Kratky and J. Shaw, edited by N. M. Klein. Karlsruhe: ZKM digital arts edition, 53–5.

Klein, N.M. 1998. Staging murders: the social imaginary, film, and the city. *Wide Angle*, 20(3), 85–96.

Klein, N.M. 2001. Absences, scripted spaces and the urban imaginary: unlikely models for the city in the twenty-first century, in *Die Stadt als Event: Zur Konstruktion Urbaner Erlebnisräume*, edited by R. Bittner. Frankfurt/M.: Campus, 450–54.

Klein, N.M. 2003. Bleeding through, in *Bleeding Through: Layers of Los Angeles 1920–1986* [DVD and Book], with additional texts by R. Comella, M. Kinder, A. Kratky and J. Shaw, edited by N.M. Klein. Karlsruhe: ZKM digital arts edition, 7–44.

Klein, N.M. 2008. *The History of Forgetting: Los Angeles and the Erasure of Memory*. 2nd Edition. New York: Verso.

Kunst als Protest. „Lasst den Scheiß!". *Die Zeit* 46 [Online, November 7]. Available at: www.zeit.de/2009/46/Kuenstlermanifest [accessed January 10, 2010].

Landow, G.P. 2006. *Hypertext 3.0: Critical Theory and New Media in an Era of Globalization: Critical Theory and New Media in a Global Era*. 2nd Edition. Baltimore: Johns Hopkins University Press.

Landow, G.P. 1997. *Hypertext 2.0: The Convergence of Contemporary Critical Theory and Technology*. Baltimore: Johns Hopkins University Press.

Lobsien, E. 1992. Großstadterfahrung und die Ästhetik des Strudelns, in *Die Großstadt als "Text,"* edited by M. Smuda. Munich: Fink, 183–98.

Marback, R. 1998. Detroit and the closed fist: toward a theory of material rhetoric. *Rhetoric Review*, 17, 74–92.

McGann, J. 2001. *Radiant Textuality: Literature After the World Wide Web*. New York: Palgrave Macmillan.

Murphet, J. 2001. *Literature and Race in Los Angeles*. Cambridge and New York: Cambridge University Press.

Ofner, A., and Siefen, C., (eds.) 2008. *Los Angeles: Eine Stadt im Film/A City on Film: Eine Retrospektive der Viennale und des Österreichischen Filmmuseums, 5. Oktober bis 5. November 2008*. Marburg: Schüren.

Scott, A.J., and Soja, E.W., (eds.) 1997. *The City, Los Angeles and Urban Theory at the End of the Twentieth Century*. Berkeley: University of California Press.

Short, J.R. 1996. *The Urban Order: An Introduction to Cities, Culture, and Power*. Cambridge, MA: Blackwell.

Simanowksi, R., Schäfer, J., and Gendolla, P. 2010. *Reading Moving Letters. Digital Literature in Research and Teaching. A Handbook*. Bielefeld: Transcript.

Sloane, S. 2000. *Digital Fictions: Storytelling in a Material World*. Stamford: Ablex.

Smuda, M. (ed.) 1992. *Die Großstadt als "Text."* Munich: Fink.

Smuda, M. 1992. Die Wahrnehmung der Großstadt als ästhetisches Problem des Erzählens: Narrativität und Futurismus im modernen Roman, in *Die Großstadt als "Text."* Munich: Fink, 131–81.

Soja, E.W. 1996a. *Thirdspace: Journeys to Los Angeles and Other Real-And-Imagined Places*. Oxford: Blackwell.

Soja, E.W. 1996b. Los Angeles, 1965–1992: from crisis-generated restructuring to restructuring-generated crisis, in *The City, Los Angeles and Urban Theory at the End of the Twentieth Century*, edited by E.W. Soja and A.J. Scott. Berkeley: University of California Press, 426–62.

Soja, E.W. 2000. Exopolis: the restructuring of urban form, in *Postmetropolis*. Oxford: Blackwell, 233–63.

Zapf, H. 2001. Literature as cultural ecology: notes towards a functional theory of imaginative texts with examples from American literature, in *Literary History/ Cultural History: Force-Fields and Tensions*, edited by H. Grabes. REAL 17, 85–99.

Zapf, H. 2002. *Literatur als kulturelle Ökologie: Zur kulturellen Funktion imaginativer Texte an Beispielen des amerikanischen Romans*. Tübingen: Niemeyer.

Zapf, H. 2005. Das Funktionsmodell der Literatur als kultureller Ökologie: Imaginative Texte im Spannungsfeld von Dekonstruktion und Regeneration, in *FunktionenvonLiteratur:TheoretischeGrundlagenundModellinterpretationen*, editeby M. Gymnich and A. Nünning. Trier: WVT, 55–78.

Zapf, H., (ed.) 2008. *Kulturökologie und Literatur: Beiträge zu einem transdisziplinären Paradigma der Literaturwissenschaft*. Heidelberg: Winter.

PART III
Ethnic Heritage and/or Cultural Commodification in the City

Introduction to Part III

Ethnic Heritage and/or Cultural Commodification in the City

Olaf Kaltmeier

Since the 1990s, cultural heritage has been playing an increasingly important role in post-Fordist urban cultural economy, which can be shown exemplarily in the revitalization and restoration of historic city centers in the Americas (Carrión 2000, 2001, Carrión and Hanley 2005, Hardoy and Gutman 1992, Scarpaci 2004). In the Americas cultural heritage has a genuinely urban dimension. In Latin America and the Caribbean urban spaces, and historic city centers in particular, make up almost half of the 82 UNESCO World Cultural Heritage Sites (http://whc.unesco.org/en/list/stat#s1). Functioning as nodal points of cultural flows of ideas, programs, aesthetics, and tourists, "heritage cities" (Kaltmeier 2011) are to be understood translocally rather than as local containers (Reguillo and Godoy 2005). Historic city centers stand out among (cultural) global cities, all of which are magnets for an emerging transnational service class as well as for cosmopolitan tourists, in that they are able to use their history and historic building stock, often certified by being listed on the UNESCO World Heritage List, as a means of distinction (Lash and Urry 1994).

While uses of cultural heritage since the nineteenth century foregrounded processes of nation-building and harmonizing national identities, the postmodern era has witnessed a proliferation and heterogenization of heritages that articulate themselves in postmodernist architecture, neo-historicism, and bricolage (Jencks 1977, 1980), in the current "retro waves" promoted by the culture industries (Guffey 2006), and in new forms of heritage tourism.

The particular importance of questions of cultural heritage within identity politics stems from the fact that it is usually "heritage experts" who make the decisions as to which material and immaterial cultural goods are considered worth preserving — and which are not. These decisions are highly relevant for identity politics and depend to a high degree on the decision-makers' power to use economic and technical resources as well as on their expert knowledge (García Canclini 1999). Cultural heritage therefore engages in a struggle for symbolic power; a struggle that depends on a quantitatively and qualitatively highly diverse distribution of different types of capital (economic, cultural, identitarian). Accordingly, socially marginal and subaltern groups in particular—unless they, like rural indigenous communities, dispose of a considerable identitarian capital—can contribute and intervene in these processes only to a relatively limited extent (Kingman 2004,

García Canclini 1999). In the post-Fordist cultural economy, questions of cultural conservation and authenticity are increasingly decided according to criteria of economic marketability rather than according to criteria of historic preservation. Ethnically identified cultural heritage in particular becomes an important resource for city marketing, as it serves to allow municipalities to create unique selling points of authenticity and cultural identity, on the one hand, and as, on the other hand, the diversity of heritage(s)—often based on migration—can endow a city with a cosmopolitan character.

In close conjunction with cultural heritage programs and the importance of nostalgia for the formation of identities, heritage tourism represents an emerging area of inquiry in works of transnational and transdisciplinary research (Porter 2008, Dann 1998, Nasser 2003, Timothy and Boyd 2003). Tourists belong to those "postmodern life-forms" (Bauman 1997) that have become a worldwide phenomenon only through the contraction of space and time brought forth by accelerating transport and infrastructure (Meethan 2001). Following Zygmunt Bauman (1994, 1997), they can be characterized, with regard to identity politics, by their lack of belonging and ties to the locus of their travels. An "easy" consumability and aestheticization, often connected to forms of exoticism (Chambers 2000, MacCannell 1999), instead determines the world of the tourist. While exoticism aims at experiencing the culturally alien, in the sense of a change of place, heritage tourism is directed at experiencing the historically different, in the sense of a time shift (Porter 2008: 268). In line with notions of a "pastime of past time" (Hutcheon 1988) or "histourism" (Römhild 1992), heritage tourism foregrounds an aestheticized consumption of history. Practices such as keeping distance and dealing critically with history and its continuation among contemporary societies—for instance, in the form of ongoing racism—are ignored here, as they would irritate the aesthetics of consumption. The current debate about tourism focuses on the consumption of place(s) and space alongside the consumption of time. The act of seeing is particularly central to the tourist's experience, thus turning "sites" into "sights" and subjecting places to the regime of the "tourist gaze." According to John Urry (1990, 1992), the tourist gaze can no longer be understood as an external lens. As a consequence of increasing global interconnectivity and mobility, it is instead subjected to translocal negotiation processes acted out between tourists and the local population, among others. This concept regards the local population not as a passive target of tourism projects, but instead as social actors who pursue their own cultural, political and economic purposes (Toyota 2006, Gotham 2007).

Using the Ecuadorian capital of San Francisco de Quito as a paradigmatic example of the restoration and revitalization of historic city centers in the Americas, the urbanists Fernando Carrión Mena and Manuel Dammert Guardia investigate the field of, often tense, interaction between the preservation and commodification of World Cultural Heritage.

Nina Möllers discusses the commodification of multiculturalism in New Orleans. She hereby focuses on the construction of Creole identities as well as the political and economic uses of ethnicity from the foundation of the city until the debates on the post-Katrina rebuilding of New Orleans.

Taking the example of the Mexican city of Mérida, Ricardo López Santillán explores how members of the indigenous group of the Maya strategically employ ethnicity to gain economic and political participation.

Literary scholar Alicia Menéndez Tarrazo (University of Oviedo) discusses the ambivalent ways in which marketing strategies of the Canadian metropolis of Vancouver and the "unofficial" discourses of the city's cultural producers and neighborhood movements both use discourses of ethnic diversity.

Works Cited

Carrión, F., (ed.) 2000. *Desarrollo cultural y gestión en centros históricos.* Quito: FLACSO.

Carrión, F., (ed.) 2001. *Centros históricos de América Latina y el Caribe.* Quito: FLACSO.

Carrión, F. and Hanley, L. (eds.) 2005. *Regeneración y revitalización urbana en las Américas: hacia un estado estable.* Quito: FLASCO.

Chambers, E. 2000. *Native Tours: The Anthropology of Travel and Tourism.* Prospect Heights: Waveland Press.

Dann, Graham M.S. 1998. There's no business like old business: tourism, the nostalgia business of the future, in *Global Tourism*, edited by W. Theobald. 2nd Edition. Oxford: Butterworth–Heineman, 29–43.

García Canclini, N. 1999. Los usos sociales del patrimonio cultural, in *Patrimonio etnológico. Nuevas perspectivas de estudio*, edited by E. Aguilar Criado. Sevilla: Instituto Andaluz de Patrimonio Histórico.

Gotham, K. 2007. *Authentic New Orleans. Tourism, Culture, and Race in the Big Easy.* New York: New York University Press.

Guffey, E. 2006. *Retro: The Culture of Revival.* London: Reaktion Books.

Hardoy, J. and Gutman, M. 1992. *Impacto de la urbanización en los centros históricos de Iberoamérica: tendencias y perspectivas.* Madrid: Editorial Mapfre.

Hutcheon, L. 1988. *A Poetics of Postmodernism: History, Theory, Fiction.* New York and London: Routledge.

Jencks, Ch. 1977. *The Language of Post-Modern Architecture.* London: Academy Ed.

Jencks, Ch. 1980. *Post-Modern Classicism: The New Synthesis.* London: Architectural Design.

Kaltmeier, O. 2011. Historic City Centers in Globalization Processes: Cultural Heritage, Urban Renewal, and Postcolonial Memories, in *EthniCities. Metropolitan Cultures and Ethnic Identities in the Americas*, edited by J. Gurr, O. Kaltmeier, and Martin Butler. Trier: WVT.

Kingman, E. 2004. Patrimonio, políticas de la memoria e institucionalización de la cultura. *Iconos*, 20, 26–34.

Lash, S. and Urry, J. 1994. *Economies of Signs and Space*. London, Thousand Oaks, and New Delhi: Sage.

MacCannell, D. 1999. *The Tourist: A New Theory of the Leisure Class*. Berkeley: University of California Press.

Meethan, K. 2001. *Tourism in Global Society. Place, Culture, Consumption.* Basingstoke and New York: Palgrave.

Nasser, N. 2003. Planning for Urban Heritage Places: Reconciling Conservation, Tourism and Sustainable Development. *Journal of Planning Literature*, 17(4), 467–79.

Porter, B. 2008. Heritage tourism: conflicting identities in the modern world, in *The Ashgate Research Companion to Heritage and Identity*, edited by H. Graham and P. Howard. Aldershot: Ashgate, 267–281.

Reguillo, R. and Godoy, M. (eds.) 2005. *Ciudades translocales: espacios, flujos, representación*. Guadalajara: ITESO.

Römhild, R. 1992. Histourismus: zur Kritik der Idyllisierung, in *Reisen und Alltag: Beiträge zur kulturwissenschaftliche Tourismusforschung*, edited by D. Kramer and R. Lutz. Frankfurt/M.: Universität Frankfurt, Institut für Kulturanthropologie, 121–30.

Scarpaci, J. 2004. *"Plazas" and "Barrios:" Heritage Tourism and Globalization in the Latin American Centro Historico*. Tucson: University of Arizona Press.

Timothy, D. and Boyd, S. 2003. *Heritage Tourism*. Harlow: Prentice Hall.

Toyota, M. 2006. Consuming images: young female Japanese tourists in Bali, in *Tourism, Consumption and Representation*, edited by K. Meethan, A. Anderson and S. Miles. Wallingford: CABI, 158–77.

Urry, J. 1990. *The Tourist Gaze: Leisure and Travel in Contemporary Societies*. London: Sage.

Urry, J. 1992. The tourist gaze revisited. *American Behavioral Scientist*, 36(2), 172–86.

Chapter 9

Quito's Historic Center: Heritage of Humanity or of the Market?

Fernando Carrión Mena and Manuel Dammert Guardia

In 1978 UNESCO declared the historic center of Quito a World Cultural Heritage of Humanity, thus recognizing its cultural and historic value, in particular the fact that it represents the largest historic city center in Latin America. Since 1988 Quito's historic center has become the object of a politics of renovation, which both local and international heritage experts have praised as an outstanding example of successful urban management, especially for the participation of public and private, national and international actors. These characteristics regularly appear as paradigmatic in the academic reflection and public discourses on historic city centers in Latin America. As this condition has been taken for granted, there has been little critical analysis or evaluation of this development to date.

This positive view suits the municipality of Quito, as it thus achieves its goal of rendering visible its own activity in a favorable manner, and because this assessment from outside allows the municipality to pursue a discourse of city marketing that positions the municipality well in international circuits. This further enables the international development cooperation to justify its actions and presence in the city as well as—thanks to (albeit limited) economic support, indebted to institutional promotion—to present potential local and national clients with the successful results of its work here.

Nonetheless, certain issues remain topical, some of which are central to the orientation of this chapter. These include, for example, the questions, whether Quito really represents a successful case of urban intervention, and, if the answer is positive, for whom it is successful. Moreover, how do the politics of renovation fit in the broader panorama of managing and organizing the city? What are the major transformations and effects they have produced? What are the characteristics of the heritage discourse that informs the above-mentioned interventions? In other words: we seek to instigate a debate about the "model of the city" that the politics of renovation in the historic center promote.

The condition of this "successful case" has recently begun to be questioned. Critics point out that the intervention comes to fruition at the margin of the reflection on a project of the city based on the belief that historic centers are almost self-sufficient. Heritage experts and intellectuals in particular perceive the historic center as an ensemble of monuments that carries a memory they need to preserve. Here, history serves to construct a stereotyping spectacle, in which heritage acts as scenario and political discourse for the legitimization of a specific type of hegemony. A further point of critique concerns the fact that parallel institutions have emerged which have imposed a political logics which sublimates the logics of tourism, the attraction of private capital, and the impact of the real estate sector, all of which taken together end up socially and economically "polluting" the urban discourses and imaginaries as well as have helped "clear" Quito of its lower social strata under the pretext of generating economic resources and restoring the city.

Monuments or Social Relationships

In the theoretical definition of the historic center, one can clearly distinguish between two concepts. The first of these is a traditional one that understands the historic center from a reductionist, unilateral, and ideological perspective that is highly charged with a notion of the past as manifesting itself through monuments. The second view seeks to go beyond the first one, making a qualitative leap in its understanding of the historic center, insofar as it moves away from the meaning of monuments as emblems in favor of a perspective in which social relationships define the determinable quality of the historic center's existence (cf. Carrión 1987).

From the perspective of a politics of intervention, the first concept has a marked weight within the notion of conservation and the treatment of the object—the historic city center—as element in its own right that is congruent with the definition of the monument. This implies a specific understanding of renovation, which Dora Arízaga defines in the following terms: "The fundamental goal of conserving the quality of values and the responsibility to leave the object of conservation to the future generation in the same condition in which we received it" (2002)[1]. The operation thus consists of "freezing" history in a particular monument in order to hand it over to future generations exactly as one has received it. This leads to certain politics of intervention and investment, as Arízaga affirms: "The public investment in the initial conservation processes will demand strong subsidized investments to stimulate the callings of the site as cause of the urban synergies that generate employment, rent, and attractions for private investment" (2002). The Quito Cultural Heritage Rescue Fund (*Fondo de Salvamento del Patrimonio Cultural de Quito*, FONSAL)[2] shares this vision with results that we will assess later in this chapter.

---

1 All Spanish citations in this paper were translated by Luisa Ellermeier, Astrid Haas, and Olaf Kaltmeier.

2 FONSAL is an institution created in 1987, after the earthquake that affected the central zone of the city. Currently, in addition to funds coming from the international development cooperation, FONSAL receives six per cent of the income tax collected by the Municipality of the District of Quito (*Municipio del Distrito Metropolitano de Quito*,

A second critical point of view concerns the definition of "historic" city centers—as constitutive elements of the city—and the necessity to move away from a "monumentalist" conceptualization, which sees the city as ensemble of monuments that is constructed on features of architectonical value, toward a conception that is principally based on social relationships. If one goes beyond the monumentalist[3] notion, the city as a whole can be understood as a historic product in each of its parts as well as in its entirety. Therefore, the entire city is historic, as are all centralities. It is from this understanding that the concept of the historic center may refer to a relationship that, in the first place, is born out of the center's central position, because the concept of the center is precisely a relationship that is constructed—in urban terms—through the concentration of central functions and through their interaction with their respective surroundings. In the second place, this relationship concerns the notion of antiquity, that is, the sum of the values of the past, which is what allows one to understand the relationship between city and historic center in its development over time. In other words, going beyond the monumentalist conceptualization means to conceive of this relationship between center and surrounding city as historically emerging from its changing social conditions rather than from material conservation.

This second concept thus has generated a critical view toward these politics that have to regard the "early" incorporation of the historic center in the city as a problem of urban planning and of the "models" of public management (cf. Cifuentes 2008). These include, for example, the modalities of financing, based on the creation of institutions that intervene by means of their "own" public resources (FONSAL) or with the help of the international development cooperation (Inter-American Development Bank, Junta de Andalucía) (cf. Rojas 2004, Samaniego 2007), the process of relocating the informal sector (cf. Valdivieso 2007), or the heritage politics and discourses (cf. Kingman 2004, Kingman and Goetschel 2005, Salgado 2008), among other aspects.

Throughout its history, the debate on the character of historic city centers had its rising and falling tides, but, owing to the impact of the earthquake that hit Quito in 1987, a significant choice was made, when FONSAL assumed a central interventionist position and managed to pursue a hegemonic politics along various axes based on a monumentalist vision of the historic city center in Quito: Firstly, FONSAL established the relative autonomy of the historic center with regard to the city as a whole, as a result of which the center ceded to nurture the project of the city and began to lose its fundamental condition: functionality. In this way, the historic center tended to be seen as being on the margin of the city and its planning

MDMQ). It is worth noting that, although a large part of its projects are related to Quito's historic center, FONSAL's activities are not limited to this area. (cf. FONSAL 2009).

3   The Diccionario de la Real Academia Española defines "monument" as "an established public work [of art] such as a statue, an inscription or a sepulchre put up in remembrance of a heroic action or any other singular event. A building that possesses artistic, archeological, historical, etc. value" (http://buscon.rae.es/draeI/).

proposal. Thereby, the centrality lost its urban condition and is now understood as being outside the relationships that constitute it, and to be perceived through the rigid frame of the "proper" attributes of monumentalism.

Secondly, FONSAL imposed the monumentalist interpretive concept and its conservationist notion of heritage politics, taking the condition of buildings as main point of reference and functionalizing historiography, stipulated by historians coming from the local elites, for that purpose.

Thirdly, FONSAL has established a model of managing the centrality with the support of national resources. Although these have an impact on real estate prices and certain economic sectors (trade, tourism, real estate), they have, in fact, not generated a principle of returning the investments made via local taxes (property tax, contributions to improvements) or national ones (VAT, income tax), which is to say that these investments have functioned as subsidies to private capital. Moreover, the international credits they command have generated project dynamics more profitable than those of urban planning, on the one hand, and the establishment of parallel administrative units that operate with the logics of the private sector, on the other hand.

In the fourth place, part of the central objectives that had been formulated was directed toward creating general conditions for attracting private capital. However, this has never happened but, instead, produced a significant depopulation, above all with regard to the lower social sectors.

This focus is to be the beginning of an erratic process that will finally result in the questioning of the condition of the "successful model." Therefore we consider it necessary to ask questions with regard to the paradigmatic condition of the "Project Historic Center of Quito" that reach across the discourses, the vision of the city that is being promoted, and the actions unfolding from the effects the project has brought forth.

This debate is obviously not limited to Quito, as it is situated in the larger context of the renovation processes of historic city centers throughout Latin America. A key issue concerns the problem of housing, which includes a discussion on gentrification processes, changes in the profile of residents, and new conditions of using space (cf. Smith 2002, Slater 2006; for an example of the incorporation of the gentrification debate in Buenos Aires cf. Herzer 2008). Critical scholars in Latin America seek to establish the topic of housing as key element for supposedly repopulating the center—without success, though, as depopulation in renovated historic city centers continues to be high.

In this context one can witness conflicts about the uses of urban space that, in the case of Quito, express themselves in an ongoing process of hollowing out public uses of urban space and in the shift from a residential to a mixed use of the historic city center for tourism and trade that have profoundly spoiled the center and, therefore, its position as "living center."

Moreover, the model of administration and management has come under scrutiny. In Quito, one has witnessed the emergence of parallel institutions that would establish themselves in private form in the Quito Historic Center Company

(*Empresa del Centro Histórico de Quito*). An increasing number of propositions demanding public rebuilding wherever results have not been satisfactory question this model. Hegemonic discourses about authenticity, memory, and the creation of meaning, among other issues recently interrogated, frame all of these interventions. This process of intervening in the historic center of Quito has been going on for more than three decades; yet to this day it has never been subjected to a real assessment that goes beyond the respective value judgments embedded in current ideologies rather than in the real processes.

The Historic Center and the City

Quito, like many important cities in the world, turned its back on its historical origins in one of its oldest spaces, the historic center: the birthplace of the city in terms of the framework of its Spanish colonial foundation. This historical disregard was impelled by a larger process of denial that began in the midst of periods of hastened urbanization (cf. Kingman 2006), first at the beginning and later in the middle of the twentieth century. These changes spurred three unprecedented developments: a striking urban expansion, which serves to distinguish the sectors of the city, separating the downtown area from the outskirts; a correlative differentiation between the old city and the modern one; and, finally, the physical and symbolic abandonment of the historic center. As the Quito Urban Development Company (*Empresa de Desarrollo Urbano de Quito*) states:

> The real estate dynamics Quito experienced during the oil period were absent from the historic center. There, contrary to the rest of the city, property deteriorated inexorably. The center began to house a sizable part of rural migrants. The crowded conditions, lack of basic services, and old age of buildings lowered their value and all forms of heritage began to erode. The quality of the environment also decreased remarkably. In those conditions, the historic center lost its functional significance, that is, it stopped being the link between the north and south of the city and became an obstacle. ... Its historic significance was also reduced: nobody identified any longer with this dirty, deteriorated, badly smelling historic center, which became a sort of example of bad urban practices and occupation of space. This situation lasted more than three decades. (INNOVAR 2008: 36–7).

This text demonstrates some of the elements of the so-called deterioration from the clearly elitist perspective of municipal policymakers: the indigenous peasant migration lowered the economic and heritage values of the site, whereby Quito's historic center lost its central functionality, and no one would identify with the center due to its supposed dirtiness, deterioration, and foul smell.

These unusual developments lead us to propose the need to identify the origin and development of Quito's historic center as part of the urbanization process.

The first point of concern is that this process began with the real and symbolic withdrawal of Quito's elite from the historic center, giving rise to the social replacement phenomenon caused by the influx of poorer inhabitants and to the territorial stigma of being a space belonging to the world of the popular classes.

Also of note is the opposition between modernity and antiquity, which comes from two historical shifts. First is a change of the city's pattern of urbanization. At the beginning of the twentieth century, the city shifted from a pattern of self-centered and low-growth development to one of high urban expansion (renovation-expansion), multi-centeredness, and it witnessed a significant population growth due to migration. The second shift concerns the establishment of the import substitution model during the 1970s that led to a metropolitan development of the city, to different kinds of centralities, and to an unparalleled growth of the outskirts, all of which resulted in the displacement of both the wealthy populations and of central functions from the city center.

For that reason, the emergence of the historic center as a real object occurred when this "crisis of centrality" arose, a predicament initially linked to the comparison between the new and the old city[4] and to the conflict between the traditional and the modern. According to this model, the "new, modern, progressive city" turns its back on the historical origin of the city by abandoning it politically, symbolically, and residentially, thus giving rise to a very powerful "urban imaginary:" the negation of the historical origin of the city or the so-called "urban parricide" (Sonenshein 1994: 142). In the case of Quito, this crisis of the historic center became so strong that the local elites considered its "rescue" or its "re-conquest" necessary.

However, since the inception of these developments and especially during the past twenty years, the opposite occurred: Quito's elites no longer deny the existence of the historic center but, in cooperation with the heritage technocrats who serve them, produce a separation of Quito's historic center from the rest of the city. Thus, the public policies in the historic center ignore the existence of the larger city in an attempt to turn the former into an enclave or a "bubble" independent from the latter. The central functions become "liquid," the population is forced out as a result of the high costs of this location, alleys are built exclusively for the purposes of luxury tourism, the historical center by day is totally different from the one by night, its accessibility is ever more restricted, and it is turning into a closed neighborhood rather than a functional city center.

---

4   For that reason, many cities of the region became defined by this comparison between the new and the old with a name that addressed antiquity such as Old Havana, the Colonial Center of Quito, or the Old City in Montevideo, among others.

The Historic Center Is the Public Space

Those urban policies that want to present alternatives in order to face the urban crisis have to deal with the character of space: Should space rather be public or private?

From our perspective, the historic center is the public space *par excellence*. As evidenced by its status of heritage of humanity, the particular and abundant legislation, the historical development of Quito evolving from its squares and streets to structure the location of population and urban activities, the whole of the historic center is greater than the sum of its parts. Additionally, the claim to public space is further supported by the confluence of the symbiotic (encounter), the symbolic (identity) and the polis (civic) at the site. It exemplifies the notion of "common place," as Quito's historic center exhibits a particular institutional framework with specific public policies, prominently including regulation and investment.

In keeping with these aspects, a project for the historic center of Quito should be socially collective, the more as the center is a heritage of humanity, engenders social identities in persons beyond the zone, and its condition of centrality is not its own, but that of the larger city. Hence, such a project should transcend time and space, while functioning within a clear framework of social confrontation, as public space is the principle milieu of the conflict between various projects, for example, the preservation that restores value to a good by returning it to its original state and the renovation that seeks to gain the value of history by stressing the past. As Françoise Choay (2007) argues, the antiquity value finally excludes the novelty value and thus threatens both the use value and the historic value. However, our proposal also encompasses the struggle about what constitutes the most desirable use of land in the historical center: Should it be residential or rather commercial? This implies considering the existence of specific interests that serve to define the economic character of the historic center.

A further important conflict about the "popular" character of Quito's historic center exists. For example, certain sectors try to stay at this site by pursuing a politics of housing (material heritage).[5] This contrasts with an understanding of this space as the environment of "popular culture" (non-material heritage)—a dimension of what is subaltern in the symbolic economy of the city. These two positions are further opposed to the official discourses and practices of history, heritage, and use of space.

The conflicts surrounding Quito's historic center also epitomize many of the symbolic elements from the various social groups in search of their "ideal" city and the construction of an "urban identity" based on it. Hence, the generation of an "urban image" is in accord with urbanization actions encompassed by a

---

5   This statement is more clearly if one considers that housing—as in no other place in the city—is highly linked to labor activities, services, stores, and equipment to the point that, if one of them is modified, the entire network changes. Thus, for instance, the relocation of street merchants involves a change in housing.

heritage discourse acting as a dispositive of power that organizes public affairs (cf. Kingman and Goetschel 2005).

It is in this context that the urban imaginaries which precede the process of urban space reproduction revaluate the importance of the historic center within the city of Quito (cf. Dammert 2009), as long as they are part of the spatial organization that occupies a prominent place in its function as "mobile frontier:" Regarding its morphology, Quito's historic center functions as an urban centrality that concentrates central functions and as a geographic one that separates and integrates the northern and southern parts of the city. In this sense, it is the meeting point of several geographically distant realities: north/south and center/periphery. Regarding time, the historic center concentrates various urban myths and imaginaries that dispute past, present, and future. It is a space that symbolically represents the frontier between the past created through memory and the future created by desire. From the perspective of social space, Quito's historic center operates as the privileged place of dispute between the public and the private.

Urban renewal, as an action that generates historical value, allows this triple frontier—characteristic of the historical center of Quito—to increase its meaning due to the contradictory process of producing inclusion and exclusion via the prioritization of certain uses of land over others (residential vs. commercial) and the promotion of narratives of (conflictual) identity. Likewise, this is not an attempt to deny the past and even less sublimate the existence of an "ideal past." On the contrary, moving away from "romantic" or "nostalgic" visions assuming an ideal past, popular and even "democratic" policies have involved a transformation of public space that goes hand in hand with the construction of a representation of meanings of Quito from *quiteñidad* ("Quitoness") in culture and the colonial character in history that, introduced by city marketing as an urban strategy and policy, assume the role of postcard narratives.[6]

Heritage: Subjects, Discourses, and Market

The intervention in the historic center of Quito comes from an urban imagery that embodies some hegemonic discursive elements, settled in specific heritage subjects, such as Quito's City Council, the international development cooperation, and the media. This discourse is based on three main components: the historical sense of what is or is not colonial, the cultural character of *mestizaje* (being *Quiteño* and being Spanish), and the social construction of a space where the popular elements can be found only as immaterial heritage. Thus, the "recovery" (of what has been lost) and the "re-conquest" (the return to the colony) become

---

6   According to a recent survey on the quality of living in Quito, 34 per cent of the interviewees indicated that "the historic center and the monuments" are what best represents the city, followed by the "tradition, art and culture" with 19 per cent (*Corporación Instituto de la Ciudad* 2008).

meaningful as ways to construct memory, but also to counter the need to create a symbolic economy of the memory that renders the real estate, tourist, and commercial interests profitable.

In order to understand the accounts of the heritage intervention in Quito's historic center, one has to consider three important elements. Firstly, the precept that heritage does not exist "inherently." It arises as the result of a complex interaction among agents who (arbitrarily) select natural and cultural guidelines employed in heritage discourses. The latter naturalize these guidelines, concealing their own production and selection process. This process, which Llorenc Prats calls "heritage activation" (1997), leads to the conceptualization of heritage as a social and historical construction and not as a "cultural tradition" (Mantecón 2005, Prats 1997) and "natural expression" of certain social groups. On the contrary, heritage, like discourses and politics, is produced and legitimized through certain agents by promoting a particular discourse about culture (identity), society (popular), and urban history (colonial).

Secondly, one must consider, the manner in which the highly political character of heritage works as a dispositive (cf. Kingman and Goetschel 2005) regulating both demographic arrangement and symbolic resignification within the exclusionary field of power. As Prats (2008) asks: To whom and what for could the activation of heritage be of interest? In principle, it is of interest to power, because without power there is no heritage (we mean the different kind of powers, but basically the political power at all its levels, and at the national level in the first place). For the case at hand, the heritage of the historic center of Quito does not rely upon a proposal "on the national level" but, on the contrary, pursues to uncover the particular features of the "local identity" in a manner similar to the regeneration policies carried out in the Ecuadorian city of Guayaquil (cf. Andrade 2006, 2007).

Thirdly, it is necessary to point out the relation between heritage and the globalization process, at least in two main contexts. The first of these is the concern and pressure exerted by international organizations and related to the recovery and preservation of the cultural heritage (tangible and intangible), all of which is expressed in the legislation, statements, and participation of certain external agents in the design of urban renovation and heritage policies (cf. Carrión 2001, Carrión and Dammert 2010, Carrión and Hanley 2005). The second context concerns the growing importance of tourism as the economic logic driving the heritage activation processes, namely, the explicit pursuit of turning social spaces into heritage sites in order to convert them into tourist attractions, which also influences the development of pro-heritage discourses (cf. Mezquita 2010). As Llorenc Prats and Agustín Santana assert (2005), we need to come to terms with the fact that tourism is not the only deciding factor in the logics of heritage politics, but it operates on the side of heritage activation (representation system), in which heritage can be "sold" and turned into a commodity. This implies:

> the contradiction, the schizophrenia that seems to exist between the local and the global: the global space gets syncretically incorporated into the *lived experience*,

while the local spaces, perceived as *authentically* local, are represented in order to be sold (although in another sense, they could also be *lived* and correspond to diverse ways of use: economic, social and ideological. (Prats and Santana 2005: 17, emphasis in the original).

The Three Axes of Intervention in the Historic City Center

In general terms, the renovation strategies for the historic city center in Quito have resulted in a broad range of policies and consequences for the area, which can be related to three principal areas: street trading, tourism, and the real estate sector.

Street Trading

The most visible restoration of the historic city center culminated in the formalization of street trading, which was taken as a demonstrative sign that the historic city center had been recovered (cf. Bromley 1998, Middleton 2003, 2009). Several social institutions and the Municipality of Quito participated in various aspects of this process, which spanned the years from 1998 until 2003. The endeavor ended with the relocation of about ten thousand street vendors to nearby shopping centers (6,000) or their dispersal to other sectors of the city (4,000).[7]

There is little information about the consequences of this process for the affected economic sector and for the historic city center. Nevertheless, Valdivieso (2007) demonstrates that in 2005 about 205 of the shops of the newly built low-priced shopping centers, where most of the merchants were concentrated, had been closed or were being used in other ways (for example as warehouses). This means that more than one thousand shops had closed. Thus the relocation of street vendors had led to a loss of the city center's commercial function.

Furthermore, the efficacy of the "restoration" of the public space has also been called into question. The premise that the public space had been "privatized" due to the use for and appropriation by informal commerce is tenuous at best. The logical outcome of a successful restoration effort would have been the full recovery of the area, marked by the respective zones' attainment of their former dimensions as public spaces after the commercial reorganization. Yet, the question remains as to whether this has been the result of the Quito project. In order to answer

---

7   In 1998, there were about 8,000 merchants, who were organized in 96 associations. They were concentrated in 22 *manzanas* (street blocks), 80 per cent of which were located in the sector of Ipiales. Eighty-five percent of these vendors were located in the "public space" and 15 per cent in shopping centers. It was calculated that the clientele served by these merchants amounted to 320,000 people, 76 per cent of them from other sectors of the city (42.5 per cent from the southern parts of Quito, 30.6 per cent from the north, and 3.7 per cent from other districts). Generally speaking, the consumers came from the lower socio-economic strata (cf. Valdivieso 2007).

this question, one must examine the basis of the overarching historic city center project, a venture which seeks to generate a clean and ordered space, one without any conflicts. Its culmination should be a space that has been converted into a tourist attraction and that serves to convey a "particular" narrative about Quito. It was with these goals in mind that the "tourist walking routes" of the historic city center were constructed, conduits which lead to and connect the *plazas* and monuments with the highest patrimonial values.

Therefore, we propose to understand the question of the restoration of the public space in terms of a dispute between the street vendors, on the one hand, and the tourist enterprises and real estate agencies, on the other. The latter benefit from the reconstruction of the urban landscape in the historic city center and, especially, from the new tourist trails. These developments were made possible by a different use of the land, namely the relocation of street trading and the new location of prestigious shops, restaurants, and hotels. This clearance of the urban and social space was accompanied by a simultaneous surge of a postcard-like narrative of the city, a formalized scene of a spectacularization of history for tourist ends.

Tourism

Boosting tourism is an explicit aim of the politics of renovation. Therefore, a heritage discourse emerged that was structured along the lines of strategic city marketing. This heritage discourse is sustained by the aforementioned trio: *quiteñidad* (identity), coloniality (historicity), and social clearance (socio-economy). From this perspective, a whole array of public interventions were developed which seek to construct an *ad-hoc* space that is delimited by the façades, the relocation of street trading, the design of exclusive walking routes, and the encouragement of investments in the tourist infrastructure.

The importance of the historic city center as a node of attraction is unquestionable. In the past decades, Ecuador has faced a constant growth of the tourist sector (both domestic and international), for which the historic city center of Quito is one of the most important attractions. According to the data of the Metropolitan Corporation for Tourism, of the recreational tourists (approximately 41 per cent of all tourists), 71 per cent have visited the historic center. Nevertheless, although the historic city center is one of the main tourist attractions of Quito, it is a zone where the tourists only stay about five hours a day on average, due to the low level of infrastructure.[8]

We can confirm the concentration of heritage-related interventions in specific areas such as the principal centers (*plazas*), certain city districts (*barrios*), determined

---

8   According to the tourist survey (*Catastro turístico* 2008), the historic center consists of the following tourist infrastructure: 25 cafés (one of the first class), six apartment hotels, five hotels (one of the first class), 15 apartment hostels (two of the first class), four hostels (two of the first class), five bed and breakfast places (one of the first class), 65 restaurants (five of the first class), and nine bars, among others.

axes of transportation (*calles*), and specific sites (namely monuments) in Quito. In this context, we would like to propose the presence of a segregated politics of intervention, one which promotes "walking routes" through particular zones while disregarding other areas, which are then eliminated from the historic landscape.

Instead of an "integral" policy of the historic city center that considers the entire area to be the spatial scale of intervention, the interventions are guided by a principle of selection, focusing on isolated and aseptic "bubbles." Obviously, this logic goes hand in hand with the hegemonic modalities of tourism, as these bubbles in the city center are designed for a specific niche in the tourist market that itself transforms the entire city center, driven by the impact of franchises and businesses, by cosmopolitan patterns of consumption, and by the flow of tourists who have no relation to the place. Tourists are given preference while local residents lose their right to the city.

The Real Estate Sector

A further explicit aim of the Municipality of Quito is the rehabilitation of the historic city center from a monumentalist perspective. Here, the real estate sector is the principal beneficiary. The annual investment in the last eight years amounted to approximately 41 million US $ on average. The overwhelming majority of funds was directed to the reconstruction of buildings (residences,[9] churches), infrastructural improvements (drains, transport), and the improvement of public spaces, all of them in some ways related to economic interests.

In the context of urban renewal, the residential question is especially important. The supply of available housing and the rehabilitation of existing houses have increased with the municipal residential projects, although this still cannot be considered as an integral housing policy. Instead, the municipality has regarded gentrification as a solution to the housing problem. On the one hand, several initiatives aimed to promote the "return" of persons of the middle- and upper-middle classes into this area. On the other hand, in the last twenty years, Quito has faced an estimated loss of 41 per cent of the total population of the historic center due to high rent and housing prices. In other words, the discourse of the "living historic center" resulted in an exodus of residents and a crisis of the promotion of the residential character of the historic city center. In 1974, the area of the historic city center was home to 90,000 inhabitants; in 2001 there were only, 51,000 residents. If this trend continues at its current rate, we can estimate that the number of urban dwellers in this area will be reduced to 15,812 persons in 2025.

---

9   One of the most important programs is *Pon a punto tu casa*, a credit program for the restoration of old houses. Between 2003 and 2007, 136 credits were granted, which financed the renovation of 88 pieces of property—including 352 apartments—with a public investment of more that two million US $ and a private contribution by the owners of less than one million US $.

Thus, the question arises as to what kind of historic city center is considered to be a "living center."

Conclusion

In the following final reflections, we would like to highlight some ideas for the understanding of the recent dynamics of historic city centers. One important notion is related to the condition of historic city centers as public spaces. We argue that public spaces are spaces that cannot be conceived of without conflict, as discussions, negotiations, and tensions form part of public life. This approach implies moving beyond those technocratic, monumentalist, and conservationist visions that separate history from social and technical interventions and thus depoliticize heritage. In order to debunk these perspectives, we argue that with regard to heritage we are facing a multifaceted issue that encompasses debates between the past and present, on the one hand, and the desired future, on the other, between social actors and economic interests, between practices and imaginaries.

These processes can be conceived of in terms of a conflictive struggle of inclusion/exclusion in the historic city center, which begins with two dynamics. First, we must mention the prioritization of specific patterns of land use, the promotion of particular narratives, as well as the deployment of heritage dispositives which aim to reduce the ensuing tensions, producing an urban aesthetics and landscape supposedly free of conflicts. The latter thus represent a veritable identity politics that condenses the connotations and significations of the historic center. This results in a negation of the possibility of generating an inclusive space that encompasses the central node as well as its surroundings, the tourist circuits as well as the popular residential zones, the richness of the historic monuments as well as the social poverty around them.

In the historic city center of Quito, these developments have generated a high level of urban segregation. Upon closer inspection one can observe a specific pattern of segregation that does not merely divide the space according to its usage or the social stratification of its users. Instead, we face a multiple fragmentation of the historic center based on different patterns of land use (economy, tourism), activities and practices, and the location of those residents who do not belong to the postcard narrative. The areas where the latter tend to live and/or work are often represented as "problem" zones in the urban maps of the city planers,[10] including spaces declared as "no-go" areas for tourists.

---

10   In the *Plan Especial para el Centro Histórico* (MDMQ 2003), the following "problem zones" are listed: a) the García Moreno prison and the neighborhood of San Roque; 2) *El Tejar, Ipiales*, and *La Merced*; 3) the "Terminal Terrestre" bus terminal; 4) Av. 24 de Mayo; 5) Av. Pichincha, La Marín. Apart from this recognition, the plan lists only a few measures that could be implemented in these areas. Therefore, these areas have maintained the status of "problem zones" over the last years.

The red light district in La Cantera in the neighborhood of San Roque, one of the most precarious and impoverished sectors of the historic center, may be one of the most representative examples of this phenomenon: the combination of civic fragmentation, the creation of an artificial image, and the lack of an integral policy. Furthermore, it is also a fragmented area in chronological terms, a space with multiple temporal landscapes. While there is a massive and never-ending movement of people and traffic in the morning, the same area at night seems to consist of an empty space, where the contrast of dark streets and bright lights form a desolate landscape, which is not easily reconcilable with the hoped-for "constructed heritage" of the area.[11]

The historic center of Quito is the stage of a strange conflict between a proposal of gentrification based on the construction of infrastructure and high-cost housing, thus aiming to generate a transformation of the socio-economic composition of the urban population, on the one hand, and the ongoing process of the *boutiquization* of the center, on the other. The latter entails little more than the changes in land use from housing to commercial purposes such as hotels, fashion and luxury goods stores, restaurants, and monuments used for personal memories (weddings) or cultural events.

A second conclusion derives from the necessity of rethinking the historic center of Quito in terms of a public space. This is especially the case *vis-à-vis* the state reforms initiated in the 1990s and the process of decentralization they have imposed. In the context of this process, competences and resources were directed to the municipalities, and many public services were privatized. One of the many consequences of this process was an ambivalent situation, whereby, on the one hand, the municipalities have been strengthened in relation to the national government, whereas, on the other hand, they have become weaker with regard to the city, as the urban government depends more on market forces than on public policies today (cf. Hiernaux and González 2008).

In this context, the administration of Quito's historic center has been decentralized, following an agreement between the National Institute of Cultural Heritage (*Instituto Nacional de Patrimonio Cultural*) and the municipality, which affirms the responsibility of the latter. On the basis of this agreement, the municipality established a private–public model of administration in which the Quito Historic Center Company—with resources from an international credit granted by the Inter-American Development Bank and FONSAL that provides national funding—generated a market-related, business-oriented logic of intervention.

The main symbol of public power, the Ecuadorian Presidential Palace, is still located at the *Plaza de la Independencia*. Nevertheless, rumors circulate in Quito

---

11   Ultimately, the situation at night changes due to the establishment of restaurants and bars in the recently restored street of La Ronda. Nonetheless, these features do not represent a transformation of the general use of the space of the historic center, but only a selective one.

of its relocation to the outskirts of the historic city center. In terms of identity politics, this could mean a devastating strike against the principal instance of democratic representation.

Indeed, Quito is facing a context in which the market appears as the object of desire for public politics, although public investment is nearly inexistent and private capital does not seem to take root in the historic center. This constellation recalls the typical contradictions of an expanding urban neoliberalism. The dismantling of the state, following the logics of privatization, leads to a deregulation that subordinates everything under the primacy of the market. While public investments are directed towards private interests, the need to create ideal conditions for attracting private capital arises. FONSAL's investments can be ultimately considered as a public subsidization of private ones (cf. Arízaga 2002). Moreover, local business people are represented on the board of directors of the Quito Historic Center Company as well as involved in decisions about public investment. Under these conditions, it seems necessary to redefine and reinforce the symbiotic, symbolic, and public factors that qualify an urban space like the historic city center.

Translated from Spanish by Luisa Ellermeier, Astrid Haas, and Olaf Kaltmeier.

Works Cited

Andrade, X. 2007. La domesticación de los urbanitas en el Guayaquil contemporáneo. *ICONOS: Revista de Ciencias Sociales*, 27, 51–64.

Andrade, X. 2006. Más ciudad, menos ciudadanía: renovación urbana y aniquilación del espacio público en Guayaquil. *Ecuador Debate*, 68, 161–97.

Arízaga, D. 2002. Proceso de financiamiento de proyectos de conservación urbana, in *Gestión del patrimonio cultural integrado*, edited by M. Zancheti. Recife: CECI.

Bromley, R. 1998. Informal commerce: expansion and exclusion in the historic centre of the Latin American city. *International Journal of Urban and Regional Research*, 22(2), 245–63.

Carrión, F. 1987. *Quito: Crisis y política urbana*. Quito: Ciudad.

Carrión, F. (ed.) 2001. *Centros históricos de América Latina y el Caribe*. Quito: FLACSO Ecuador.

Carrión, F. and Dammert Guardia, M. (eds.) 2010. *Quito: ¿Una metrópoli mundial?* Quito: OLACCHI.

Carrión, F. and Hanley, L. (eds.) 2005. *Regeneración y revitalización en las Américas: hacia un estado estable*. Quito: FLACSO Ecuador

Cifuentes, C. 2008. La planificación de las áreas patrimoniales de Quito. *Centro–h*, 1, 101–14.

Corporación Instituto de la Ciudad. 2009. *Quito, un caleidoscopio de percepciones: Midiendo la calidad de vida*. Quito: MDMQ.

Choay, F. 2007. *Alegoría del patrimonio*. Barcelona: Editorial Gustavo Gili.

Dammert Guardia, M. 2009. Patrimonio y producción del espacio en las políticas de renovación del Centro Histórico de Quito. *Argumentos: Revista de análisis social* 3(2) [Online]. Available at: http://www.revistargumentos.org.pe/index. php?fp_verpub=true&idpub=170 [accessed: July 27, 2010].

Fondo de Salvamento del Patrimonio Cultural de Quito (FONSAL). 2009. *Quito, patrimonio y vida: Obra del FONSAL 2001–2008.* Quito: FONSAL.

Herzer, H. (ed.) 2008. *Con el corazón mirando al sur: Transformaciones en el sur de la ciudad de Buenos Aires.* Buenos Aires: Espacio Editorial.

Hiernaux, D. and Gonzáles, C. 2008. ¿Regulación o desregulación? De las políticas sobre los centros históricos. *Centro–h*, 1, 40–50.

INNOVAR. 2008. *Memorias de una transición: De empresa de desarrollo del centro histórico a INNOVAR.IUO.* Quito: MDMQ.

Kingman, E. 2004. Patrimonio, políticas de la memoria e institucionalización de la cultura. *ICONOS: Revista de Ciencias Sociales*, 20, 26–34.

Kingman, E. 2006. *La ciudad y los otros: Quito 1860–1940: Higienismo, ornato y policía.* Quito: FLACSO Ecuador.

Kingman, E. and Goetschel, A.M. 2005. El patrimonio como dispositivo disciplinario y la canalización de la memoria: una lectura histórica desde los Andes, in *Regeneración y revitalización urbana en las Américas: Hacia un estado estable*, edited by F. Carrión and L. Hanley. Quito: FLACSO Ecuador.

Mantecón, A.R. 2005. Las disputas por el patrimonio: transformaciones analíticas y contextuales de la problemática patrimonial en México, in *La antropología urbana en México*, edited by N. García Canclini. México, D.F.: Fondo de Cultura Económica.

Mezquita, A. 2010. El turismo en la ciudad de Quito, in *Quito: ¿Metrópoli mundial?*, edited by F. Carrión and M. Dammert Guardia. Quito: OLACCHI–INNOVAR.

Middleton, A. 2003. Informal traders and planners in the regeneration of historic city centers: the case of Quito, Ecuador. *Progress in Planning*, 59(2), 71–123.

Middleton, A. 2009. Comerciantes informales en la rehabilitación de los centros históricos: el caso de Quito, Ecuador. *Centro–h* 3, 5.

Municipio del Distrito Metropolitano de Quito (MDMQ). 2003. *Centro Histórico de Quito: Plan especial.* Quito: MDMQ.

Prats, Ll. 1997. *Antropología y patrimonio.* Barcelona: Ariel.

Prats, Ll. and Santana, A. 2005 (eds.) *El encuentro del turismo con el patrimonio cultural: Concepciones teóricas y modelos de aplicación.* Sevilla: Fundación El Monte.

Rojas, E. 2004. *Volver al centro: La recuperación de áreas urbanas centrales.* New York: Inter-American Development Bank.

Salgado, M. 2008. El patrimonio cultural como narrativa totalizadora y técnica de gubernamentalidad. *Centro–h*, 1, 13–25.

Samaniego, P. 2007. Financiamiento de centros históricos: el caso del centro histórico de Quito, in *El financiamiento de los Centros Históricos de América*

Latina y el Caribe, edited by F. Carrión. Quito: FLACSO–INNOVAR / Lincoln Institute.

Slater, T. 2006. The eviction of critical perspectives from gentrification research. *International Journal of Urban and Regional Research*, 30(4), 737–57.

Smith, N. 2002. New globalism, new urbanism: gentrification as a global urban strategy. *Antipode*, 34(3), 434–57.

Sonenshein, R.J. *Politics in Black and White: Race and Power in Los Angeles*. Princeton: Princeton University Press, 1994.

Valdivieso, N. 2007. *Modernización del comercio informal en el Centro Histórico de Quito* [Online]. Available at: http://www.reseau-amerique-latine.fr/ceisal-bruxelles/URB/URB-1-VALDIVIESO.pdf [accessed: July 27, 2010].

Chapter 10
"Economically, We Sit on a Cultural Gold Mine": Commodified Multiculturalism and Identity Politics in New Orleans

Nina Möllers

"Crescent City," "Big Easy," and the "City that Care Forgot"—New Orleans is known by many names, and both political administrations and the tourism industry know how to capitalize on the city's reputation as fun-filled, exciting, and unique place. As one of the top ten most visited cities in the United States, New Orleans's economy heavily relies on tourism revenues. Visitors come for different reasons, but they all have certain images and ideas about the city at the Gulf of Mexico which may or may not correlate with reality. Like some of its nicknames suggest, the "City that Care Forgot" has been known best for its spectacular, at times unbridled nightlife as well as its rich musical heritage and delicious exotic food. An every bit as important unique selling proposition is New Orleans's diverse ethnocultural history, attributing to its image as the most African or northernmost Caribbean City (Kemp 1997). Its multicultural heritage has both fascinated and at times frightened visitors for more than two hundred years. Though "its history, full of romantic incident and legendary lore is in itself sufficient to fill a volume" (Jewell 1874), this chapter focuses on the role of New Orleans's ethnically diverse culture, its utilization in current tourism marketing efforts, and the concurrent (de)construction of ethnic identities, specifically the Creole[1] identity. Starting out with a brief historical retrospect on the origins of New Orleans's diversity, the chapter will then discuss the present commodification of multiculturalism in New Orleans's tourism industry, particularly after Hurricane Katrina, and the debate over the "essence" of racial and ethnocultural identities and identity politics within this context.

---

1   "Creole" denotes the descendants of the original French and Spanish settlers coming from Europe in the eighteenth century as well as those immigrant families coming in after 1803 that resembled the original *ancienne population* in ethnocultural heritage, religion, and language.

Creating the "Gumbo:" Identity and Ethnicity in New Orleans

New Orleans's history has always been multiethnic. Founded by the Canadian Jean Baptiste Le Moyne de Bienville on behalf of the French Mississippi Company in 1718 and named after the French Duc d'Orléans, the swampy city at the mouth of the Mississippi soon became a new home for settlers from diverse European, Caribbean, and (though mostly not voluntarily) African regions. After the Seven Years' War, the vast Louisiana Territory that reached from Canada in the North to the present-day states of Montana, Wyoming, Colorado, and Oklahoma in the West was ceded to the Spanish kingdom in the Peace Treaty of 1763. After a temporary return under French control in 1800, Napoleon sold the Louisiana Territory including the city of New Orleans to the young American republic in the Louisiana Purchase of 1803. Due to its ethnic multiculturalism, New Orleans has often been characterized as rather "un-American." From language, architecture, music, and food to religion, judiciary and political culture, "NOLA" has been signified as the "American Other" (Möllers 2008: 53–73). With its mixture of European, African, and Caribbean influences, New Orleans has created an image of itself as the ethnic "gumbo," named after the famous Creole stew dish. In fact, ethnic diversity is *the* constitutive factor of New Orleans's identity as a city that is extensively used, reproduced, and performed by its gigantic tourism industry. Capitalizing not only on its multicultural heritage, but also playing with its image as "un-American," "mythic," and quite simply "unique" on the North American continent has become an absolute imperative. The city's tourism marketing corporation also jumps on this bandwagon, when it comes to characterizing New Orleans on its official website:

> Normally when tourists or first-time residents come to New Orleans, they have a difficult time understanding the city. It looks like no other place in the United States. ... But it is an American city—just a very different place with a very peculiar history. New Orleans is a place where Africans, both slave and free, and American Indians shared their cultures and intermingled with European settlers. (Hirsch and Logsdon 2010)

The peculiarity of New Orleans's culture and heritage as alluded to in this text written by two eminent historians at the University of New Orleans, already had an enormous appeal with visitors in the nineteenth century. After it became "American" with the Louisiana Purchase in 1803, many reporters and literates traveled the newly acquired region eager to decode the city's mythic nature. For them, the "Otherness" of the Gulf city was particularly "embodied" in the many multiethnic individuals going about their daily business in the crowded streets of the French Quarter. Feeling caught in a limbo between Europe, Africa, and the Caribbean, the New Yorker bookseller and travel reporter Will H. Coleman recorded his bewilderment at the sight of this ethnic "hodge-podge" in the following way:

A man might here study the world. Every race that the world boasts is here, and a good many races that are nowhere else. The strangest and most complicated mixture of Negro and Caucasian blood, with Negroes washed white, and white men that mulattoes would scorn to claim as of their own particular hybrid. … The air is broken by every language—English, French, Italian and German, varied by gombo languages of every shade; languages whose whole vocabulary embraces but a few dozen words, the major part of which is expressive, emphatic and terrific oaths. (Coleman 1998: 89, 91)[2]

Especially contributing to the ambiguous and irritating ethnic and racial mixture were the Free People of Color—or Creoles of Color—who had occupied the middle space in the three-tiered social system during the French and Spanish colonial period between a dominant white society of Creole and American background, on the one hand, and a large black slave population, on the other. The Latin colonial population's laxer attitudes toward interracial sexual relations, the lack of white marriageable women, relatively generous Spanish manumission laws, and a high number of immigrants in the wake of the Haitian Revolution had resulted in a large group of Free People of Color, making up 18 per cent of the entire free population in Louisiana in 1810. In New Orleans, Free People of Color even accounted for over 30 per cent of the free population in 1805 (Cummings 1968: 45, Hanger 1996: 57). Often well-trained and educated, the Free People of Color constituted a distinctive economic, social, and cultural class in the multicultural community of colonial and antebellum New Orleans. French-speaking, Catholic, wealthy, and in some instances slaveholders, these Creoles of Color identified with the white Creole Latin population.

In contrast to other free colored communities, Louisiana's Free People of Color enjoyed several important rights: they were allowed property, they could bring suit even against whites, and they could bear arms. However, being free, wealthy, and "racially-mixed," they were also perceived as a potentially threatening anomaly in the binary American system and its ideology of essentialized race identities based on the one-drop-rule. Fearing the destabilization of their slave society, the Americans therefore worked towards clear-cut definitions of ethnic and racial identities and restricted the social and economic power of racially-mixed "in-betweens." Closely connected to the exceptional political, economic, and social status of the Free People of Color, the identification marker "Creole" became the most contested identification feature in Louisiana. Laden with representational power to create, sustain, and erase identities and thereby decide over in- or exclusion into collective identity communities, the category "Creole" was politicized and commodified. Taking part in the "Creole identity"—in its consumption as commodified signifier, as property—allowed for having at least a share in the hegemonic social status.

---

2   See also Bremer 1854: 213, Featherstonhaugh 1844: 140, Latrobe 1951: 22, Stuart 1833: 198.

The resulting processes of negotiation and redefinition were played out against the background of a smoldering culture conflict between the white Latin *ancienne population* and American newcomers. Over the first three decades of the nineteenth century, both groups battled over influence in politics, law, economy, and society. Supported by the strengthening of the American nation after the War of 1812 and the economic security unfolding in its wake, the scales were soon tipped for the Americans. Although the Latin population was now forced to work *with* the Americans instead of against them, differences in language, social habits, and culture prevailed and were cultivated as "Creole peculiarities," adding spice to the image of Louisiana and specifically New Orleans as the "American Orient."

Reconstruction politics following the Civil War then offered the Creoles of Color a unique opportunity for fighting for full political and social integration into the American nation. Literate, well-educated, wealthy, and with an impressive history of military action to show for themselves (Möllers 2008: 88–98, 191–217, McConnell 1968, Hollandsworth 1995), they began their political crusade by establishing newspapers, organizing citizens' committees, and aggressively pushing the Union city government towards granting them franchise in upcoming elections. "Creole" as an identity marker played an important role in this endeavor by emphasizing their conformity with the white Creole culture while simultaneously de-emphasizing their difference, supposedly visible in their partly black racial heritage. Soon, however, this strategy began to show deficiencies. In their fight for a restructured and more equalitarian Southern society, the Colored Creoles were caught between their culturally constructed identity as Creoles and their increasingly racialized identity as African Americans. In the fight for political and social integration into the American nation, the identity marker "Creole" was continuously re-negotiated within a semantic field and an arena of political action that was influenced, at times constrained, by white definitions of "Creole," on the one hand, and reactions to this construction by the Anglo African American community, on the other.

Even so, its contested nature has not resulted in meaninglessness. Creole culture and identification continue to play a distinctive role within the multicultural world of New Orleans. Disregarding the historical complexities of identities and ethnicities in New Orleans, however, Creoles are nowadays often used—and at times reduced—to serve as "model-minority" and template for postcolonial ideas of Creolization and postmodern multicultural identity politics. Tying back to the concept of "Creoleness" as an identificatory "third space in-between," created out of the mixture of formerly hierarchically structured cultural influences, New Orleans as a city capitalizes on this specific Creole heritage as a powerful signifier for ethnic diversity and multiculturalism and performs it in manifold "cultural attractions."

Commodifying Creoleness and Multiculturalism in New Orleans in the Twentieth Century

Utilizing cultural diversity as marketing tool in tourism is of course anything but new. A few cursory glances at tourism guidebooks of the nineteenth century suffice to see that cultural diversity and otherness have always been used as enticement for tourism. For cities like New Orleans, luring visitors to come and spend money in restaurants, cultural institutions, and at festivals is essential for survival. The product that New Orleans sells is its "unique culture" incomparable to anything elsewhere. From the food they eat over the houses they sleep in to the music they hear and the handicraft they buy as souvenirs, visitors are indoctrinated with the impression that all of this is 1) directly derived from the city's "authentic" multicultural heritage and 2) inimitable. The enormous economic potential connected to such a cultural "branding" did not go unnoticed for long, even in New Orleans, which is not particularly known for progressive, cutting-edge business ideas. Lieutenant Governor Mitch Landrieu, having headed Louisiana's Department of Culture, Recreation, and Tourism (DCRT) before assuming New Orleans's mayoralty in May 2010, initiated a "Roadmap for Change" in which the mission to be accomplished was formulated as: "Position Louisiana to Lead Through Action in Defining a New South Through Culture, Recreation and Tourism" (Louisiana Office of Lt. Gov. 2005). By conducting interviews with stakeholders in business, tourism, academia, research, culture, and the arts, the DCRT brought to the forefront current perceptions of Louisiana and New Orleans and asked for recommendations for strengthening the region's and city's touristic attractiveness. The necessity to take advantage of the region's unique cultural assets, especially its ethnocultural diversity, was the consensus:

> Focus on the culture of the place and what can be economically viable with Louisiana's culture. … Support the recognition that Louisiana's culture can drive a new economic engine—arts and culture—to build tourism. … Take advantage of Louisiana's unique "foreign" flair and promote it. (Louisiana Office of Lt. Gov. 2005)

On the municipal level, the Bring New Orleans Back Commission (BNOBC), appointed by former mayor C. Ray Nagin in 2005, was in line with this approach when proclaiming:

> Economically, we sit on a cultural gold mine. A long-term, sustained investment in our culture is the most viable way to yield a first-class city. Culture is the identity of New Orleans. When we conduct business with the soul, spirit and intelligence that our culture demands, we will experience unimaginable economic growth. (Cultural Subcommittee 2005)

The commitment to branding and commodifying New Orleans as unique cultural experience was particularly pressing after Hurricanes Katrina and Rita in 2005. In the wake of these devastating storms, tourism as the second largest economic pillar of Louisiana, slumped due to damaged infrastructure, escalating crime, and complete shut-down of many tourist attractions. Just weeks following Katrina, Lieutenant Governor Landrieu and his staff developed a strategic plan called "Louisiana Rebirth: Restoring the Soul of America," which in every of its four points explicitly referenced to multiculturalism as a key to a fast and sustainable recovery of Louisiana and New Orleans. By focusing on multicultural aspects within the cultural economy—both in terms of supply and demand—the DCRT aimed at rebuilding New Orleans as the quintessential place for multicultural tourism in the United States. Acknowledging the importance to "promote Louisiana as a multicultural destination that is inclusive of all races," it contracted an advertising company "owned, operated, and staffed by people of color to create and implement marketing strategies designed to capitalize on the burgeoning African American and Latino travel markets" (Louisiana Office of Lt. Gov. 2005a). A clear definition of the concept of multiculturalism or even a discussion of its complex and entangled history in New Orleans apparently was not deemed necessary. Exploiting on its supposed economic assets, multiculturalism simply served as a rather empty vessel for "everything that sells:"

> Some of the marketing efforts designed to attract *multicultural* first-time visitors and reinforce Louisiana as a top destination among repeat visitors include print and website advertising, *multicultural* festival sponsorships such as Essence, mall tours, in partnership with *multicultural* publications, *multicultural* sporting events held in Louisiana, newsletters, web promotions, and research projects that query *multicultural* travelers about their travel and spending habits. (Louisiana Office of Lt. Gov. 2005a, emphasis mine)

A similarly crude sledgehammer method was used by the BNOBC when proposing the following steps in order to regain strength and bring New Orleans back on its feet:

> Lets [sic] aggressively promote a New Orleans with less prejudice by embracing the truly integrated nature of our culture. ... People come from around the world to see us live and play. We need to strengthen our culture for internal consumption. Lets [sic] encourage our population, in its entirety, to cook our food, to dance our dances, play our music, respect our architecture, and tell our stories. Lets [sic] support our culture in tangible ways that ensure an organic and self-sustaining development. From the food served at City Hall to the music played while callers are on hold, we should utilize every opportunity to embrace our own distinctive way of being. (Cultural Subcommittee 2005)

This statement is meaningful in several ways. First, it celebrates consumption as the panacea for the survival of both New Orleans's unique culture and its economy. By consuming their culture through architecture, food, music, and the like, New Orleanians foster their cultural collective identity and, almost incidentally as it seems, capitalize on their economic "gold mine." Culture is no longer seen as the manifestation of a particular identity, heritage, lifestyle, or set of values, but rather as spirit and purpose to which to aspire in order to secure wealth and satisfaction. In a way it seems as if New Orleanians first had to be persuaded themselves—to be culturally "re-educated," so to speak. The above-noted quotation is further very telling because it displays the great tension, even discrepancy, between the claim to celebrate diversity, on the one hand, and the apparent need for conjuring a feeling of togetherness and collective culture, on the other. Repeating the mantra "Diversity is a strength, not a weakness" (Louisiana Office of Lt. Gov. 2005a) in all public appearances and printed material, Landrieu aimed at putting Louisiana's unique multicultural heritage at the center of a hopefully revenue-producing cultural economy that expended little effort to define multiculturalism, let alone such basic concepts as culture or identity. At the bottom line, culture means business (as one of the DCRT's reports so tellingly is called) and therefore needs to come in easily accessible and sellable compartments, not fluid, creolized, or contested terrains of negotiation. "Louisiana's cultural economy," defined as "[t]he people, enterprises, and communities that transform cultural skills, knowledge, and ideas into economically productive goods, services, and places" (Louisiana Office of Lt. Gov. 2005b: 6), is a supposedly authentic, uncontested, essential, and *sui generis* untapped cash cow:

> Louisiana has an economic asset that other states can only dream of: a multifaceted, deeply-rooted, authentic, and unique culture. ... As a state that has long relied on oil, gas, and timber to fuel its economy, Louisiana is now realizing that, in culture, it may have a new source of largely untapped economic energy. (Louisiana Office of Lt. Gov. 2005b: 4)

In all of this, New Orleans has basically ceased to be a city and has instead become a brand. This "multicultural brand New Orleans" makes recourse to images, concepts, and wishes that may not necessarily find equivalents in reality, but which exquisitely create the indispensable "metaphorical story" that creates the "emotional context people need to locate themselves in a larger experience" (Webber 1997: 96). In a very superficial, but extremely powerful and economically successful way, this branding discourse is enriched with bits and pieces of postcolonial theory to legitimize the construction of a multicultural and yet in its diversity miraculously homogeneous culture—if we believe the advertising brochures. Borrowing from postcolonial concepts such as "third space" and "transculturation," branding experts, politicians, and tourism agents construct an image of New Orleans as unified in its diversity and thereby safe to be consumed. The originally subversive concept of third space "marked by the fusion of cultural elements ... permanently 'translated'" is re-essentialized as the one and only cultural space available—the "authentic" New Orleans waiting for the visitor to be enjoyed (Hall 2003: 30–31).

At the Heart of It All: What is Creole?

The extensive use of words and concepts such as "unique" and "authenticity" is of course highly problematic and does not go uncontested by New Orleanians. In fact, the rather shallow use of multiculturalism in the context of cultural economy and tourism has reheated smoldering conflicts such as the "Creole controversy" dating back to the 1980s. Counteracting tendencies in the tourism industry and popular perceptions of Louisiana culture—both heavily interrelated—that primarily equate Louisiana's unique cultural heritage with Cajun music and food, Creoles of diverse ethnocultural make-up worked against being subsumed under the blown-up Cajun label and started to actively identify as Creoles in the 1990s. Their newly found self-confidence resulted in the formation of several interest groups and institutions such as C.R.E.O.L.E. Inc. and the Un-Cajun Committee fighting for public visibility and the survival of their culture (Ayres 1997).

This so-called "Creole Renaissance," which began as a cultural awareness and recognition movement, soon also tackled more fundamental questions about politics and economy. Aside from the everlasting debate about the racial composition of "the Creole identity," Creoles felt politically unrepresented as an ethnic group and economically disadvantaged in their effort to sustain their culture. Especially after Hurricane Katrina, which showcased many of Louisiana's grievances, Creoles felt neglected in their particular needs for recovery. The partially sensational media coverage in the wake of the hurricane at times even operated against them, for instance, when the renowned restaurant critic Alan Richman suspected that Creoles "are a faerie folk, like leprechauns, rather than an indigenous race" (2006). The most authoritative and substantial response to this unqualified statement, which triggered a lively, at times aggressive debate, was voiced by the Creole Heritage Center, situated at the Northwestern State University in Natchitoches, which initiated a petition calling for full political and cultural recognition of Creoles as an ethnic group:

> Since the 2005 hurricanes, recent national media representatives have questioned the survival of this culture and even its actual existence. ... By signing this petition, we call for the following: That the Creole culture is not only accepted and its existence acknowledged, but that our legislators and educators take steps to include this important culture in the history books. That support is given to the Creole Center in its continuing pursuit of providing research and documentation of the national Creole culture, which has produced the only National Creole Family History Database of its kind. That the Creole culture and heritage, rarely acknowledged in spite of its uniqueness, is worthy and deserving of attention

and preservation; without it an important part of the American experience could be lost (Petition 2008).

In the many mostly supportive responses to the petition, Creoles affirmed their cultural and ethnic identity as "authentic Creole," although there are many discrepancies when it comes to defining this identity marker. In their affirmations many Creoles fall back on essentializing argumentations themselves, often not even shying away from utilizing highly problematic terms:

> My Creole heritage is authentic, undeniable, and 100% American. … Although many of us are transplanted, and labeled under different racial/cultural groups, we are here. Further, I … would appreciate it's [sic] recognition and preservation for our children, who are also of Creole blood. (Llorens 2008)

While tensions between Creoles of Color and African Americans had temporarily been covered up by the efforts of the Civil Rights Movement to speak with one voice, the "Creole Renaissance" reawakened the debate about the racial meaning of Creole identity. In an effort of reconciliation, the Creole Heritage Center supports a racially-inclusive or rather racially undetermined definition of Louisiana Creoleness:

> Creoles are not one thing or the other, and have lived their lives being misunderstood, misrepresented, and misinterpreted. In the past, under White government, Creoles were not allowed to be an equal part of society. Blacks, free and slaves, did not feel Creoles were part of their world either. Because of this rejection, Creoles had a strong bond with one another and had to create their own world and culture. (Creole Heritage Center 2006)

Surely, there is much to say on the formulation of this definition, including the question of agency in the construction of Creole identity. What becomes clear, though, is that "Creole" is seen as something in-between—a third space beyond the established categories, mainly those of "racial" identity—and offering the promise of inclusion into a multicultural society while maintaining self-identifying powers. The process of translating different cultural elements into something new allows for the active "filling" of the identity container instead of just reacting to existing forms of demarcation or simply reproducing them. It is this sense of conscious activity and the embrace of an affirmative attitude that sets this definition apart from historical ones and aims at fulfilling the postcolonial and postmodern postulate of fluid, permeable, and self-determined identities.

Soon Creoles even went a step further when they started campaigning for the inclusion of the category "Creole" into the 2010 Census forms. Supporters of this agenda argue that their multiethnic and multicultural background cannot be reduced to one of the existing ethnocultural categories. They are no longer willing to see themselves as "Other," but rather want to identify themselves as an integral part of the American nation while actively and positively embracing all the different pieces of their identity, including its "racial mixture." The emphasis on their multiracial heritage, however, also leads to conflicts with members of the larger African American community, who suspect elitist and exclusionary tendencies behind the Creoles' renewed effort to revitalize their collective and—from parts of the African American community's point of view— evasive cultural heritage.

The fact that the political recognition of a "Creole" identity would have economic consequences further stirs the contention. Since Louisiana has no clear numbers on their Creole and Cajun communities, it is regularly shut out from its share in 400 billion dollars of annual federal funding for cultural and educational programs. Aside from racial background, the 2010 Census for the first time includes a question about cultural origins. Under question 8, asking about Hispanic, Latino, or Spanish origin, Louisiana Creoles are prompted to check "Another Hispanic, Latino, or Spanish origin" and then add "Louisiana Creole" or "Cajun" in the field below. Despite its apparent inoffensiveness, this advice propagated by many Creole activists' groups has created quite a controversy. In order to be counted as an ethnic community comparable to Native American tribes, Census officials ask the Creoles to identify in this specific, uniform way. Simply identifying as "Creole" without the specification "Louisiana Creole" would result in a non-count because this could mean almost anything from Haiti Creole to French Creole. Some activists revolt against this drive for uniformity, however, and fear a "misrepresentation" or even miscount as "Latino:"

> I would love more cultural federal funding as well as recognition as an ethnic group but reporting as Latino or as a Creole race would not be accurate. Until there is a way to report your ethnicity (such as Irish or Italian) I will continue to report my race as Black and non-Hispanic. (Anonymous 2010)

Yet, for many others, subsuming "Creole" under "Black" constitutes the real misrepresentation. Calling Creoles "Black" also elicited a strong-worded response from A.D. Powell, an activist of the interracial movement, when I first started researching the Louisiana Creoles' history. In contrast to the above-quoted anonymous commentator, Powell sees Creoles much more accurately described along the lines of Hispanics:

> Creoles are a multiracial ethnic group, just as Hispanics are. Hispanics are not called "black," even though nearly all of them have African ancestry. Creoles should be respected in the same way. Creoles, like Hispanics, vary in phenotype from European or "white" to African or "black," and every shade and phenotype in between. ... Calling them a "black" group when so many of them are as "white" as you are, is as racist as calling them "non-Aryan." (Powell 2003)

In its vehement language, the comment points to the still present and highly problematic underlying racial dimension of Creole identity. True multiculturalism in the sense of a "third space" is not easily realized just by acknowledging Creole identity. Though comprehensive and laudable, the effort for a Creole census category carries with it challenging baggage, because for some the petition represents only another effort of a dominant-within-non-dominant group, so to speak, to gain influence to the detriment of smaller and less powerful ethnic communities. Comments such as this one by a blogger are easily to be found in diverse internet forums:

> A "Louisiana Creole" category is simply another "Hispanic/Latino" category —what about the rest of us multiracials? The Louisiana Creoles are going to get their escape hatch, while the rest of us are left out to dry? ... In reality, Louisiana Creoles make up such a small insignificant part of the US population, and someone wants to give THEM a block to check off before they give ALL mulattoes one? This is stupid. (Miller 2008, emphasis in the original)

Hence, the fight over identities is on. While Creoles—of all shades and backgrounds—campaign for the recognition as separate ethnocultural group, African Americans and whites in turn fear exclusion and misrepresentation. The Family History Database of the Creole Heritage Center claims to aim at strengthening the Creole heritage by offering the opportunity to research individual family trees. This at least evokes the question: What about those Creoles who are unable of providing a "flawless" Creole family tree back into the eighteenth century, but who feel, live, and identify as Creoles? Is the Creole identity category truly inclusive and non-essential? And where does this contested nature of identities leave us with regards to the commodification of culture?

Conclusion

The Creole self-determination movement and New Orleans's cultural economy strategies do not operate independently from each other. The tourism industry as well as the state and city governments and the Creole population itself rely on specific images of Creole culture and Creolization; mostly, however, to very different ends. The tension between the postmodern constructivist language of racial and cultural self-identification, fusion, and creolized identities in the third space stands in stark contrast to a powerful tourism machine that, to be sure, superficially builds its case upon exactly this multicultural aspect of the Creole culture, but which ultimately needs solidified, unified, and conform cultural categories to work with. And even the Creole activists, irrespective of their particular interests, reach the limits of their identity politics when truly trying to stick to postmodern identificatory and cultural fluidity. Trapped between individual self-determination and collective identity politics in which economic, cultural, and racial interests clash, simplification and usurpation are hardly avoidable.

Thus, they face the well-known dilemma between "the constructivist language that is required by academic correctness and foundationalist or essentialist message that is required if appeals to 'identify' are to be effective in practice" (Brubaker and Cooper 2000: 6). Concepts of Creolization and their subversive aspirations are, at least in this context, to be seriously doubted in their operative effectiveness. During the nineteenth century, when racial and cultural identities were mainly fixed by hegemonic institutions, creating and affirming a creolized "third identity" surely was a resistant and liberating act. In the days of postmodern fluidity and at least rhetorically granted permeable boundaries, however, Creole identity apparently has lost much of its subversive power. It has all become one big blur that even runs the risk of ultimately erasing cultural identities, making it even easier for the tourism industry to capitalize upon the rather mushy Creole heritage that is New Orleans. And when the fight for identities becomes identity politics, identity categories almost always become re-essentialized. Nations, states, city governments, but also activist groups play into the hands of homogenization and subvert their own ideals out of necessity:

> [T]he substantivization of hybridity in the form of reifying resistance—in a movement, as a (more or less) self-consciously cohering intervention—at once homogenizes the heterogeneous, fixes the flux and flow, orders the dis-orderly, renders more or less safe by "capturing" the transgressive expression of the hybrid. (Goldberg 2000: 82)

The operationalization of Creole identity as the quintessential expression of multiculturalism thus reveals many practical problems. Who will judge on the affiliation to the group and on what grounds? If there are no restrictions, the category—as the blogger suggested—will be devoid of any meaning. If certain ethno-cultural characteristics are taken as the foundation of a Creole identity, the category is turned from an inclusive into an exclusive and repressive category. In their attempt to overcome identity dichotomies, the Census activists are really supporting those skeptics who argue that too many categories result in arbitrariness and "identity-hopping." And finally, the racial component of the Creole label and its power to exclude are just as strong today as they were in the nineteenth century. As Stuart Hall has suggested, effects of past colonial mechanisms stay influential in that the axis between colonizer and colonized is simply transferred onto internal differences within the de-colonized society.

With its aim of plundering New Orleans's cultural "gold mine," city government and tourism economy do their bit to the comprehensive undermining of Creole culture in New Orleans and its replacement by a thoroughly commodified but lifeless *Ersatz*. The irony of it all is that tourism actually realizes the danger of cultural erosion, since its marketing strategies depend upon at least some "originality" and "authenticity" of culture in order to be marketable. The "erosion of the authenticity of [New Orleans's] underlying culture" is in fact detrimental to its economic value (Louisiana Office of Lt. Gov. 2005b: 65).

The debate between Creoles (of Color) and Anglo-African Americans about their share in New Orleans's particular cultural heritage is only one arena of negotiation among many that bespeaks of the intensified battle of interests between different cultural groups at the beginning of the twentieth-first century. The commodification of culture, initiated and pursued by the city government and the tourist economy under the smoke screen of multiculturalism, in reality leads to cultural turf wars and undermines the survival and future development of exactly that on which the cultural economy bases its right to exist: an "authentic" culture, a culture grown out of a region's community, performed by the people, not staged, because that is how people want to live and not because a tourism industry information leaflet tells them to.

Works Cited

Anonymous. 2010. Comment to Sturgis, S. Census Watch: in Louisiana, the census gets a dose of Cajun pride: *Facing South: Online Magazine of the Institute of the Institute for Southern Studies* [Online]. Available at: http://www.southernstudies.org/2010/03/census-watch-in-louisiana-the-census-gets-a-dose-of-cajun-pride.html [accessed: June 30, 2010].

Anonymous. 2006. *Creole Folks*. 2006 [Online, October 29]. Available at: http://creoleneworleans.typepad.com/creole_folks/2006/10/creole_folksyou.html [accessed: November 1, 2006].

Augustine/Comeaux. 2005. Letters to the editor. *French Creoles* [Online]. Available at: http://www.frenchcreoles.com/LetterToEdit.html [accessed: June 16, 2005].

Ayres, B.D. 1997. On Bayou, non-Cajuns fight for recognition. *New York Times* [Online, November 23]. Available at: http://www.nytimes.com/1997/11/23/us/on-bayou-non-cajuns-fight-for-recognition.html?pagewanted=1 [accessed: July 2, 2010].

Bremer, F. 1854. *Homes of the New World: Impressions of America*, vol. 1. New York: Harper and Brothers.

Brubaker, R. and Cooper, F. 2000. Beyond "identity." *Theory and Society*, 29(1), 1–47.

Coleman, W.H. 1998 [1885]. The french market, in *Louisiana Sojourns: Travellers' Tales and Literary Journeys*, edited by F. De Caro. Baton Rouge: Louisiana State University Press.

Creole Heritage Center. 2006. *Creole Definition* [Online: Northwestern State University]. Available at: http://www.nsula.edu/creole/definition.asp [accessed: March 27, 2006].

Cultural Subcommittee of the Mayors Commission to Bring Back New Orleans. 2005. *The Cultural Vision Statement. A Cultural Dream for New Orleans: a Symphony of Integration.* [Online]. Available at: http://www.bringneworleansback.org/

Portals/BringNewOrleansBack/portal.aspx?tabid=71 [accessed: January 11, 2006].

Cummings, J. 1968. *Negro Population in the United States, 1790–1915.* New York: Arno Press.

Dyson, M.E. 2006. *Come Hell or High Water: Hurricane Katrina and the Color of Disaster.* New York: Basic Civitas Books.

Featherstonehaugh, G.W. 1844. *Excursion through the Slave States, from Washington on the Potomac to the Frontier of Mexico; With Sketches of Popular Manners and Geological Notices.* New York: John Murray.

Goldberg, D.T. 2000. Heterogeneity and hybridity: colonial legacy, postcolonial heresy, in *Companion to Postcolonial Studies*, edited by H. Schwarz. Malden, MA: Blackwell, 72–86.

Hall, S. 2003. Créolité and the process of creolization, in *Créolité and Creolization, Documenta 11_Platform 3*, edited by O. Enwezor et al. Ostfildern and Stuttgart: Hatje Cantz, 27–41.

Hanger, K.S. 1996. Patronage, property and persistence: the emergence of a free black elite in Spanish New Orleans. *Slavery and Abolition*, 17, 44–64.

Hirsch, A.R. and Logsdon, J. 2010. *The People and Culture of New Orleans.* [Online]. Available at: http://www.neworleansonline.com/neworleans/history/people.html [accessed: July 1, 2010].

Hollandsworth, J.G., Jr. 1995. *The Louisiana Native Guards: The Black Military Experience during the Civil War.* Baton Rouge: Louisiana State University Press.

Jewell, E.L. (ed.) 1874. *Crescent City Illustrated: The Commercial, Social, Political and General History of New Orleans, Including Biographical Sketches of its Distinguished Citizens.* New Orleans: n.p.

Kemp, J.R. 1997. When the painter met the creoles. *Boston Globe* [Online, November 30]. Available at: http://www.boston.com/globe/search/stories/books/books97/christopher_benfey.htm [accessed: July 1, 2010].

Latrobe, B.H. 1951. *Impressions Respecting New Orleans: Diary and Sketches 1819–1820*, edited by S. Wilson, Jr. New York: Columbia University Press.

Llorens, A. 2008. Comments to petition [Online, July 6]. Available at: http://www.ipetitions.com/petition/creole/signatures-1.html [accessed: July 6, 2008].

Louisiana Office of Lieutenant Governor. Louisiana Department of Culture, Recreation and Tourism. 2004. *Roadmap for Change.* [Online]. Available at: http://www.crt.state.la.us/DOCUMENTARCHIVE/ [accessed: June 30, 2010].

Louisiana Office of Lieutenant Governor. Louisiana Department of Culture, Recreation and Tourism. 2005a. *Louisiana Rebirth. Restoring the Soul of America. Diversity Initiative Executive Summary Overview.* [Online]. Available at: http://www.crt.state.la.us/documentarchive/index.aspx#TrueLouisiana [accessed: July 1, 2010].

Louisiana Office of Lieutenant Governor. Louisiana Department of Culture, Recreation and Tourism, Office of Cultural Development. Louisiana Division of the Arts. 2005b. *Louisiana: Where Culture Means Business*. [Online]. Available: http://www.crt.state.la.us/DOCUMENTARCHIVE/ [accessed: June 30, 2010].

McConnell, R.C. 1968. *Negro Troops of Antebellum Louisiana: A History of the Battalion of Free Men of Color*. Baton Rouge: Louisiana State University Press.

Martel, B. 2006. Storms payback from God, Nagin says. *Washington Post* [Online, January 17]. Available at: http://www.washingtonpost.com/wp-dyn/content/article/2006/01/16/AR2006011600925.html [accessed: February 14, 2006].

Miller, R. 2008. Comments posted. *The Study of Racialism Forum* [Online, April 22, May 26]. Available at: http://www.onedroprule.org/about4628.html [accessed: June 21, 2008].

Möllers, N. 2007. *Kreolische Identität: Eine amerikanische "Rassengeschichte" zwischen Schwarz und Weiß: Die Free People of Color in New Orleans*. Bielefeld: Transcript.

NSU Creole Heritage Center. 2008. *Petition* [Online]. Available at: http://www. ipetitions.com/petition/creole/index.html [accessed: July 6, 2008].

Olmsted, F.L. 1984 [1861]. *The Cotton Kingdom: A Traveller's Observations and Cotton and Slavery in the American Slave States*, edited by A. M. Schlesinger. New York: Modern Library.

Powell, A.D. 2003. Email to Author. [accessed: September 3, 2003].

Richman, A. 2006. Yes, we're open. *GQ* [Online, November]. Available at: http://www.gq.com/food-travel/alan-richman/200611/katrina-new-orleans-food [accessed: July 2, 2010].

Stuart, J. 1833. *Three Years in North America*. 2 vols. 3rd Edition. Edinburgh: Printed for Robert Cadell.

Webber, A.M. 1997. What great brands do. *Fast Company,* 97(10) [Online]. Available at: http://www.fastcompany.com/magazine/10/bedbury.html [accessed: July 2, 2010].

Chapter 11

Mobilizing Ethnicity: Yucatecan Maya Professionals in Mérida and their Participation in the Cultural and Political Fields

Ricardo López Santillán

For the analysis presented here, I discuss processes that involve Mayan indigenous people who live in Mérida, Mexico, who have roots in underprivileged rural families, yet who have succeeded in entering the job market in positions typical of middle-class professionals: almost all of them as scholars or employees in government agencies, especially those designed to preserve culture and the recognition and promotion of their specific ethnolinguistic group. Holding these jobs has led to a kind of "marketing" of their ethnicity, where their Mayan cultural heritage is construed as symbolic capital that can result in a diversity of rewards, such as a heightened public presence or income increase. This minority of indigenous professionals hold prestigious jobs with good wages when compared to most of the Maya residing in the city who work in low-skilled positions and earn low incomes. In order to achieve these social and economic gains, many Maya have left their home communities in search of a better education and a more favorable labor market. Furthermore, I want to emphasize that these professionals have passed through a process of "acculturation" or "Mexicanization" thanks to the higher levels of formal educational and upward social mobility present in Mérida.

Mérida, the capital city of Yucatán (located on the Peninsula of Yucatán, Mexico), houses almost one half of the state's total population and centralizes most of the region's cultural, political and private sectors, resulting in a local economy with a very high rate of non-rural jobs. This concentration of wealth and social institutions (universities, transportation, medical care, housing, well-paid jobs, etc.) represents a veritable oasis in a state where abject poverty and marginality are the norm. The Maya professionals, who are at the center of this study, implement government plans and programs designed to promote their traditions and customs in a variety of public venues that have both local and international audiences. For example, they showcase their traditions locally in a festival that takes place every Sunday in Mérida's main square. Furthermore, they promote their language with Mayan language radio and television and often hold contests to write songs or short stories. In the international arena, they organize events in the United States, principally in the San Francisco Bay area, where a substantial Yucatecan Maya population has taken root (Cornelius et al. 2007) and are also active hosting celebrations such as the International Mother Language Day, sponsored by UNESCO. In addition, this work also discusses the paradox revolving around the fact that, although there has been a great increase in public awareness of Yucatecan Maya culture, this change has not eliminated erstwhile discriminatory ways of thinking from the local culture. I began my research with this idea, starting from some observations on government agencies where Maya professionals work. While some of these official matters have been documented, I realized that in order to gain a wider perspective it was imperative to search for *life stories*, that is, for a deeper and more complete analysis based also on the point of view of the main actors.

As a first consideration of this study, subjects were asked about their ethnic identity. "Ethnicity," according to Barth (1976), is inextricably based on differences between one group of people and another and is not necessarily defined by objective identifiers such as clothing, language, housing, and lifestyle values, but rather is more often based on factors that members of each group consider significant differences, such as self-identification with an ethnolinguistic group. It is important to make clear that this work relates to the population of Yucatecan Maya, also known as Peninsular Maya. This ethnic identification is mentioned in the interviews as it relates to different aspects of the subjects' lives. Despite these instances of identification, one should not forget that some subjects nonetheless use masking strategies in very specific moments in an effort to hide ethnic identifiers in order to avoid discrimination. As it stands, this work is based on twenty interviews that were conducted using the life story method, where all of the interviewees were qualified based on their origin, ethnic identity, level of education, residence in Mérida or surrounding suburbs, age (between twenty-five and sixty), forming part of the workforce, and having a monthly household income of at least approximately $5,496 MXN per capita.

It is important to briefly note that the use of life stories is a technique based on an ethnosociological principle, which means that empirical evidence taken from those individuals interviewed is used to form a database when there is no statistical evidence available. The life stories technique is used to create models based on recurring facts in the interviewees' stories. Accordingly, there is no set number of interviewees for a project; this number instead becomes clear when there is enough common evidence collected from different stories in order to speak in more general terms about a given subject (Bertraux 2005). Having several interviewees provides validity because one can note the similarities between events and facts described in order to compensate for the subjectivity of individual interviewees. Using this method, a researcher is able to validate the facts described by the subjects by identifying common points made by other interviewees from the same environment (or group, social or professional category, etc.) and, in so doing, can achieve the effect of obtaining from "many voices one story" (Resendiz 2001: 154), in this case a story of Maya professionals.

With regard to the present chapter, it must be noted that all individuals interviewed completed their education and began their professional careers in Mérida. All of them have been residents of this city, although in two cases the interviewees currently reside abroad where they are completing doctoral programs. The interviewees are from different backgrounds: four are from Mérida's suburbs, which two decades ago still were rural; in two cases the interviewees come from the neighboring state of Campeche; and the remaining interviewees come from rural areas inside the state of Yucatán.

The Yucatecan Maya in Context

According to the latest *Population and Housing Census* (INEGI 2001), the state of Oaxaca has the largest number of indigenous people in Mexico in absolute numbers, totaling over 1.5 million people and equaling 47.9 per cent of the state population. The bordering state of Chiapas has the second-highest indigenous population with a little over a million people, making up about 25.5 per cent of the total state population. The third highest population of indigenous people is in the state of Yucatán, with almost a million people (981,064); and yet Yucatán is the only state where the population of indigenous people is greater than that of non-indigenous people. According to the census, 59.2 per cent of the entire population identify themselves as part of an indigenous group. Another factor that distinguishes the indigenous population of Yucatán from that of other states is that it is by and large culturally homogenous and from a single ethnolinguistic group.[1]

According to more recent data obtained from the 2005 *Population Count* (INEGI 2006),[2] the three states do not report significant changes in population. Nonetheless, Yucatán reports a several per cent drop in the number of indigenous residents from 2000–2005. The indigenous population fell from 59.2 to 51.55 per cent, from 981.064 to 937.691 inhabitants. What explains this change? There are two possible reasons. The first is that there is a strong migratory movement to seek employment in the tourist locations of Cancún, Playa del Carmen and the Mayan Riviera in the neighboring state of Quintana Roo. Recently these three locations have seen the largest population gains in the entire country, and the majority of the

---

1   According to the census, 91.2 per cent of Yucatecan Mayans are bilingual, that is, they speak both the Yucatecan Mayan language and Spanish.

2   In Mexico, the Institute of National Statistics, Geography and Data, which has recently changed its name to the Institute of National Statistics and Geography (INEGI), is an autonomous public agency charged with conducting a Population and Housing Census every ten years. The last of these was in 2000. The Population Count is also taken every ten years, but five years after the census; this is done in order to provide more recent information. The two reports are not entirely equal, because the Population Count reports much less information, as it uses a much shorter form than the Population and Housing Census.

new residents are from the state of Yucatán and especially the Yucatecan Maya. In fact, the largest measured indigenous migration nationwide has been from Yucatán to Quinatana Roo (Gaultier 2003). The second reason has to do with the masking of ethnic identity: for underprivileged Maya lacking education, being indigenous has a stigma, while this does not apply to the same extent to Maya professionals. The former is understandable, because even though there are efforts by the government to recognize the importance of indigenous people as an integral part of the national culture—and this is particularly true of Yucatán—the Yucatecan Maya are still victims of racism and symbolic violence (López Santillán 2007). The practice of marginalization reinforces the status quo and has the effect of relegating indigenous groups to the lowest part of the socioeconomic scale.

For this reason, the common perception is that when one mentions the Yucatecan Maya, one refers to field workers planting corn in small communities in rural parts of the state, or if they are Maya from urban areas, they are imagined to be working in underpaid positions for unskilled workers: gardening and construction for men and street vending and housekeeping for women. This explains why self-identification as a member of an indigenous group holds a stigma and why people have developed strategies for hiding the fact that they belong to this undervalued ethnicity, as reflected in both the Census and Population Count. Nonetheless, given that Yucatán is the state with the largest proportion of indigenous people, and especially since they derive from the same ethnolinguistic group, it is not surprising that the Peninsular Maya, being the majority population, have a huge presence in every aspect of public life. This is true even though the dominant view, both among common citizens and among politicians and academics (sociologists, historians, ethnographers, and anthropologists), is that this group is marginalized and poverty-stricken due to its ethnic origins. This reflection persists even in light of the fact that there are important contingents of indigenous people that have risen to professional positions.

The upward social climb that the Yucatecan Maya have experienced in recent years has happened for a variety of reasons. There are structural factors, such as the implementation of the development model of "import substitution," based on the premise that Mexico should attempt to reduce its foreign dependency through the local production of industrialized products. This, along with rapid industrialization and urbanization of the country from the 1940s up until 1980, produced strong economic growth. Despite the economic crises (1982, 1994, and 2009), there have still been possibilities for social advancement, albeit increasingly fewer in number. In any case, this process of accelerated industrialization and urbanization gave rise to a "deruralization" of the countryside: rural areas stopped receiving government aid to such an extent that farming became unviable, in turn initiating migrations from the country to the city of Mérida. This structural situation led both individuals and families to decide that, because they had no future in their rural home communities, they would migrate to the city for more opportunities.

This migration from rural areas to Mérida (considering the city as a metropolitan area) started in 1950s but became extremely intense during the 1970s and once again between 2000 and 2005. Indeed, when farming in Yucatán did not provide enough income, Mérida became a magnet for people, as it was easy to find work there. Even if this work paid poorly, it still provided more than life in the fields, and people could better survive than in the country, even while living in urban poverty. This is because Mérida has a strong urban infrastructure that offers schools, health centers, durable housing, and much better sanitary conditions; in other words, the surroundings facilitate better living standards in comparison to any rural community in the state and in particular in relation to the small isolated indigenous communities of the Peninsula of Yucatán.

Mérida, of course, was not the only destination for immigrants; as noted above; recently there were those who migrated to Cancún and the Mayan Riviera in the neighboring state of Quintana Roo or a few even to the United States, However, in the case of Yucatán, those who abandoned their rural communities migrated primarily to the capital city of Mérida. Currently the interior parts of the state are very sparsely populated while Mérida and its surrounding suburbs hold around one half of the state's population, and in addition, it is the principal population center of the Yucatecan Maya on the entire peninsula (Ruz 2002). In accordance with data of the 2005 *Population Count,* Mérida has reached a total population of 897,740, that is, 49.35 per cent of the total inhabitants of Yucatán. Among them, around one third assume an ethnic identity (Serrano 2002) and 105,283 are speakers of their indigenous native language. Thus, the city of Mérida has been the scene of measured geographical mobility which in turn sparked social mobility, in particular for those Maya who had been poor in their communities of origin but who beat the socioeconomic status of their rural uneducated parents. In fact, most of the Maya intellectuals and professionals come from a disadvantaged background that was only surmountable via very important personal and family efforts and, as mentioned above, structural advantages only offered by the city of Mérida (López Santillán 2006).

It must be mentioned that obtaining education has not been easy for the majority of indigenous people in Mexico. In general, their communities are rural, spread out, and have small populations. Normally, rural community schools only offer the lower grades of elementary school or, in some cases, full elementary education up to the sixth grade. In the best of situations, a few communities provide teachers for the national "telesecondary" system of distance education, where students can take classes by watching televised courses. For this reason, if students want to continue and graduate from high school, they must migrate to a more urban environment. For high school, there are very few options in areas outside the capital. In the case of public universities, up until the last decade there were no options outside of Mérida.

It is important to note that during the entire twentieth century, education in Mexico held explicit homogenizing nationalist goals, which supplanted cultural awareness of the distinct indigenous groups to the point that, under the guise of nation building, institutions subjugated local cultures of diverse regions and ethnicities. This process of homogenization begins in elementary school but becomes more and more evident in higher grades. While bilingual and bicultural elementary schools exist, there are no bilingual universities. University courses and seminars are always taught in Spanish, even in recently created indigenous universities. Their only real difference to other universities lies in their being located in key urban centers with a strong indigenous presence. As a result of the national goal of homogenizing the country linguistically and culturally, Yucatán's institutions have met the important "benchmark" of acculturating the indigenous population. In this sense, the government has systematically turned the Maya into Spanish speakers at the cost of completely debilitating the strength and importance of their native language. In the case of Mexico, and Yucatán in particular, the fact that indigenous people have been able to become educated and become professionals means that, for the most part, they have adopted the cultural patterns and goals and even the language the Mexican government set out for them. Put in another way, it was "necessary" to "de-Indianize" or "Mexicanize" the Maya to enable them to ascend the socioeconomic ladder and be recognized in society.

Mobilizing Ethnicity

Clifford Geertz (1963) said that the two objectives that indigenous communities have in modern states are, on the one hand, to be recognized as entities whose opinion matters and, on the other hand, to be part of the mechanism of the state to effect changes in the way people live, such as being able to participate in the political structures, justice systems, and finally even to have political power on a local and even global level. All of this depends on civics: it is contingent upon the ability of groups to recognize their rights and responsibilities.

The Neo-Weberians believe that ethnic groups can use their status as capital in political and economic environments. This means that ethnicity is not just used to protect traditions and specific aspects of culture, but that it is also an element in the struggle for power and has weight in some forms of social relations (Malešević 2004). In other words, ethnic positioning implies the mobilization of an identitarian capital in order to achieve social and political recognition—what is also called "identity politics," where governments gain an image of "good governance" among others by institutionalizing minority rights (Thies and Kaltmeier 2009). Gutiérrez Chong (2007) writes that in the case of Latin America, there have been centuries of exclusion and discrimination of the national status of indigenous peoples, which has provoked them recently to mobilize nationalist movements based on their ethnicities. In this way, indigenous people become immersed in a process of defining their identity, negotiability before the state, and redefinition of the meaning of their past. The pragmatic forms of ethnic mobilization in Mexico are presented on two fronts: 1) indigenous people are more involved in the reconstruction of their past and in political participation, as is the case with the nationalization of natural resources, and in search of democratic transformations on a large scale, as is the case in Ecuador and Bolivia where the indigenous past was used to construct a national identity. 2) The other extreme is the case of Mexico, where an understanding of the indigenous past has been systematically erased and then reformulated by the government so as to be integrated into the national identity.

This consideration helps make the point that some Yucatecan Maya, in particular those with a higher level of education, have been able to utilize their capital (cultural, symbolic, identitarian, or even educational) in the so-called field of identity politics (Thies and Kaltmeier 2009) in order to legitimize themselves and thus gain presence in certain aspects of society in the sense that Bourdieu proposes (1979). In the state of Yucatán, paradoxically, the ethnic mobilization has not taken the path of separatist or autonomist movements; instead it is apparently harmonious and in line with the goals of the state institutions. Movements are both political, akin to political parties and their antiquated corporate practices (Quintal 2005, Mattiace 2007), and, as we will see here, cultural, as they work through information clearinghouses and cultural promoters from official agencies. The latter try to transform what Thies and Kaltmeier have named the "'exchange rate' between identitarian and economic capital" (2009: 35) in the sense of gaining not only legitimization but also other rewards in the social arena.

One can see then that the Mexican state, in its construction of the idea of nation, has undermined the cultural specificity of indigenous communities so as to make them conform to nationalist ideals. Nonetheless, as paradoxical as it seems, at the very moment indigenous people finish higher education and achieve some professional success their ethnicity changes into cultural capital that can benefit them. This transformation happens to a greater degree with people who work for national and state institutions charged with working directly or indirectly with indigenous populations, such as the National Institute for Indigenous Languages (INALI); the National Council for Culture and the Arts (CONACULTA), which works through the Department of Popular Culture and Indigenous Peoples; the National Council for Educational Development (CONAFE); or the National Commission for the Development of Indigenous Peoples (CDI), each of which is a federal agency with a local office. In addition to these, the agency most relevant to the region, the Institute for the Development of Mayan Culture (INDEMAYA) is an employer of many Maya professionals. Private sectors also capitalize on cultural and symbolic heritage, albeit to a lesser degree. For example, local mass media outlets such as radio and television stations offer broadcasts in Yucatecan Maya.

It must be said that the city of Mérida hosts all these public (local or federal) institutions and private business that have ethnic employees or employees who assume ethnic identity. This generates intense cultural and political dynamics that characterize the city and, from there, impact the entire state. Although these official institutions and their employees often are being dislocated to or conduct, field work or develop projects in various communities around the state, these different occurrences cannot compare to what the capital city offers. The fact remains that most of their working life and project planning as well as most of the public events organized by these groups (exhibitions in museums, craft shops, public entertainment, documentary videos, lectures, or book presentations of Maya authors) are centralized in Mérida. This cultural life, marked by Mayan folklore, concentrates in the city because there it can reach a larger population: the Indians who live in the city, as well as other non-ethnic actors, including domestic and international visitors.

It is important to reiterate that the processes this study refers to are not limited to Yucatán; nor is the current development a new phenomenon; on the contrary, this way of dealing with issues has been practiced for at least a generation and has been expanding the whole time. There is moreover a larger number of indigenous people involved in mainstream positions in many different Mexican states. In fact, sociological and anthropological literature in Mexico since the 1970s has shown that the indigenous populations are not isolated from national life. Instead, they are an integral part of the market because of their purchasing power, and they make significant contributions through their interactions with official institutions (legislative, juridical, and executive powers), even when these interactions sometimes prove to be one-sided and fruitless (Stavenhagen 1969, Pozas and Horasitas 1971, Gonzalez Casanova 1975). Nonetheless, the trends which researchers referred to then are much more evident now. Currently the state has embraced (at least in the words of the Mexican government) the notions of multiculturalism, aligning itself with international treaties on respect and recognition of indigenous groups, such as Agreement 169 of the International Labor Organization (ILO).

Again, it bears repeating that cases of upward social mobility of indigenous peoples of Mexico in general and of Yucatán in particular, even with their idiosyncrasies, are just a few of the many worldwide cases. Nonetheless, they follow basically the same principle: first there is an introspective revaluation of the structures of the state, and then an ethnic group mobilizes in order to adjust their unfavorable position within the society. In the case of Mexico, it is important to note that there are crucial reasons for addressing the inequalities and for introducing changes that help improve interethnic relations in the country: the most recent event indicating the imperative need for change was the uprising of the Zapatista Army of National Liberation (EZLN) in 1994 that brought about negotiations between the EZLN and the federal government.

In addition, the place of the individuals in and their interaction with society remain a relevant issue. Bourdieu (1979) noted that different cultural practices hold a certain symbolic power including influence, political advantage, honor, and significance. Individuals can capitalize on those factors, even economically, because all modern societies operate according to the same principles. In Mexico, and particularly in Yucatán, one can see a clear paradox in the current social dynamic that has been established between the indigenous and *mestizo*, or non-indigenous, populations. Those individuals with ethnic origins, including people who are educated professionals, are still victims of oppression and symbolic violence carried out non-indigenous groups or individuals. For this reason, the Mayan language and some indigenous customs are losing ground in the interest of national homogenization. However, it is also true that Mexican society has gradually come to recognize the contribution of native peoples within the strange concoction that is the "national culture," and there are various groups that seek to establish more consistent and harmonious relationships between indigenous and non-indigenous peoples.

The Marketing of Ethnicity

There are extreme cases of marketing ethnicity in Mexico, such as some Maya communities in Yucatán who persist in "selling" representations of traditions usually practiced in rural areas not only to tourists but also to anthropologists and ethnologists. There are those who are willing to perform the Maya ceremonies in months of the year that do not correspond to their usual season simply in order to make them fit into researchers' field work schedule or tourists' vacation periods (Lizama 2007). Perhaps the most extreme case of this phenomenon can be witnessed in the neighboring state of Quintana Roo, where some locations have become a stage for international tour operators to bring hordes of visitors to find "a genuine traditional Maya village," and where tour guides speaking Spanish and Maya simultaneously communicate with locals and visitors in a kind of spectacle arranged for those purposes (Marín 2009). I propose an analogy that I hope will be fair: the *Lederhosen* and *Dirndl* worn and the festivities of the Munich *Oktoberfest* are not at all similar to the clothing worn by most Germans today or to the way they usually party.

Though not as extreme as these cases, the cultural activities that take place in Mérida, many of them are promoted by the official institutions charged with propagating Mayan culture, are nevertheless a very relevant case in point. Those events recreate an unsubstantiated vision of indigenous cultures that completely decontextualizes and undervalues the practices and customs of indigenous peoples in an unrealistic folklorist representation. In fact, in many cultural events in Mérida, government department officials, among them some Maya professionals, exhibit stereotypes or characterizations of how the Maya "really live today." These spectacles are partly cultural and partly political marketing at the same time. In Mérida, both for people without ethnic affiliation and for local and foreign tourists, Mayan culture is represented as kind of show act. This is also the case with Mayan-language literature and song contests, traditional dances at the Plaza Mayor, where participants wear "typical" attire, the Festival of the Dead (*Hanal Pixan*), and many others. Even documentaries and educational videos in the Mayan language made by members of the Maya community often characterize the Maya in a way that ends up being a folklorist representation of reality (Duarte 2008).

In Mexico there are several institutions like the CDI, National Institute of Anthropology and History (INAH) and the National Foundation for Tourism Development (FONATUR), whose mission is part of the revaluation of the legacy of tangible cultural heritage (pyramids and other archeological sites) and the symbolic-cultural heritage of native peoples (food, clothing, housing, crafts, rituals). These institutions show a great need for indigenous professionals who become advocates and promoters of cultural tourism projects; disseminators of certain practices, including commercial projects (such as artisanship, cooking, traditional medicine) and the arts (such as dance, music, and indigenous literature); and participants in various events such as UNESCO's International Mother Language Day, among others.

In the case of the Yucatecan Maya, it is worth noting that, due to their specific cultural heritage of being indigenous, they can capitalize on their ethnicity in two ways: first, in terms of social prestige, as their cultural capital can translate into symbolic capital that gains in strength and visibility as people move up in society, and secondly, the cultural capital of ethnicity can also translate into financial capital, as ethnicity can be used for financial gains or even to achieve upward social mobility. The experiences of the Maya are incontrovertible and cover a broad spectrum of processes from geographical mobility to social advancement. One example that is pertinent to my study concerns partisan politics and/or official institutions. In the first instance, there are, as previously mentioned, Maya individuals working in government agencies directly linked to the rescue and revaluation of Mayan culture. Moreover, from the stories I have collected one can see that many of the interviewees have worked on the campaign committees of various national political parties, including those who have lead political rallies in rural communities, delivering their speeches in Maya. There are also individuals who have worked in government agencies as members of programs to support those Maya living in rural communities, whose role is to receive requests which later become government plans and programs. Among the interview subjects there is a doctor from the Mexican Social Security Institute (IMSS), who uses his language skills to make more accurate diagnoses, especially among the older Maya population, among which there is a greater degree of monolingualism or limited ability to speak Spanish. The younger respondents include some who have been instructors for CONAFE in remote communities (where the command of Spanish is still very basic). Several of these individuals obtained scholarships for higher education in local universities because of their work with CONAFE. It is worth mentioning that there are ways to capitalize on ethnicity, although to a lesser extent, in areas other than government agencies: the most prominent example in Yucatán includes scholarships for postgraduate studies offered by the Ford Foundation and intended solely for indigenous students. There are also Mayan language teachers teaching courses for students of foreign universities and/ or private courses in Mérida as well as those working in the media, broadcasting radio or television spots.

All of these stories confirm that the Maya professionals benefit their employers, and that their ethnicity has brought them rewards (symbolic and financial) as a result of their language skills and their having grown up with traditional practices of their culture such as crafts, cooking, and knowledge of the land and its flora and fauna, as well as the intangible elements of their heritage such as stories, legends, history, festivals, and rituals, among others. In addition to the professional and economic success that derives from this background, these men and women have the satisfaction of knowing that their work has been key to the preservation of their language and traditions. At the same time, they have contributed to the building of a more positive image of their ethnic group, one which is not linked to marginalization and poverty. Yet it must be said that some Maya professionals have also built symbolic barriers between themselves and the underprivileged, undereducated Maya.

In any case, being a successful indigenous person is highly significant to the interviewees, who believe that this effort by themselves and many other indigenous Maya will have positive effects, for example that the Maya in general will be neither ashamed to speak their language nor tempted to deny their origins, but will instead be able to live "with their heads raised high." Wieviorka (2003: 25) notes that, despite the current forms of discrimination, most of today's societies go through a process of reversing this negative definition of ethnic identities ("shameful identities" as they are classified) for a positive identity built on recognition. While in exceptional cases this may develop into radical communitarianism, dogmatism, or fundamentalism, this is not the case with the Yucatecan Maya. Many Maya professionals state that their ethnic origin has led to academic achievements as well as professional and financial support. To cite just two examples, one interviewee, who works in a center for anthropological research, emphasizes that she has published several of her works in Maya and a bilingual edition of traditional stories of which she has sold almost 12,000 copies. Another interviewee, an archaeologist at INAH, says that because of his ability to speak Maya he has been invited to deliver talks and presentations of his work in many national and international forums.

As already mentioned, societies tend to reverse the negative values of ethnicity. The case of the revaluation of the Yucatán Maya is related to not only to the work of official agencies but also the value of others, the non-Maya, who have contributed to the culture and the preservation of the language in recent times. This is the case of people involved in the construction of more equal and fair social relations. In Yucatán, it is increasingly common for politicians, doctors, nurses, lawyers, residents, tourists, and, of course, social scientists from various backgrounds and disciplines, to be drawn in by both the local cultural specificity of the Maya and, in a broad sense, society as a whole. Along with the Maya, they are gradually working to reduce the asymmetries of interethnic contact.

These processes serve to alleviate the symbolic violence suffered by the Maya from the colonial era, which continues up until today, despite all clamoring about inter- and multiculturalism. It can only be noted that some sectors of local society, such as the upper and upper middle classes still practice racism of low intensity on the Peninsula of Yucatán. Nonetheless, many situations of interethnic contact in Mérida have become less conflictive, in some part because the Maya recognize these issues and are able to bring about positive improvements. Recognizing the problems and learning to confront them may empower Maya professionals to influence others within their own ethnic group so as to overcome the anxiety of speaking their language in public and, someday, to finally overcome the stigma that weighs on them and often pushes them to various forms of auto-marginalization.

However, as has been suggested in this chapter, periods of ethnic mobilization in Mexico are very often guided by the official institutions and the political use of indigenous culture in ways that are deeply rooted in semi-authoritarian government systems. It is important to note that in this sense Mexico's politics have never changed significantly. The state, whether ruled by parties of the center or the right, channels power vertically, using the very branches of government, trade unions, professional associations, chambers of commerce, and finally groups of various kinds, in a manner that control never leaves the hands of government agencies and other official bodies.

According to Quintal (2005), the political participation of the Maya in Yucatán is calculated by several organizations aligned to political powers, as they have been linked to local governments for some time. It must be said, though, that these groups sought some degree of autonomy during the 1990s. However, logic in an omnipresent state, professionals move their Mayan cultural and symbolic capital to the place where it will benefit them most, and this is most always within official institutions. In this process their single objective is to pursue strategic moves that ensure public recognition of their culture and that can also benefit them economically. It is also important to note that through their political participation, Maya people show solidarity not only among members of their ethnic group but, more importantly, with people from their community and especially their family members. In fact, it is worth further investigating the workings of these networks of reciprocity that facilitate the integration of the Mayan cities and certain niche markets, and how their culture benefits from these networks in both the economic as well as political arena. However, forms of political participation among Maya professionals are very cyclical and do not necessarily imply a permanent party affiliation. Changes are common, because individuals have the possibility of obtaining more advantages by joining the party in power at a given moment. This appears to be an act of indecisiveness but is instead regarded as a strategic action. There are only a few staunch party members, activists, and those individuals who have occupied elected positions (such as the local deputies) who permanently retain their party affiliation.

Conclusion

In Yucatán, the mobilization of symbolic capital based on ethnicity is a strategic action that moves predominantly through channels of political or governmental institutions, especially those linked to the dissemination, promotion, and preservation of cultural and historical heritage of the group and that can bring rewards in the form of prestige and public recognition in addition to the income derived from employment in these institutions. There is paradox entailed in the fact that, in Mérida in particular but, by extension also in many other Mexican cities with a strong ethnic presence, indigenous populations face a strong pressure to "Mexicanize." Once they have turned into successful urban professionals (having graduated and being established in a professional position) these assimilated indigenous populations are rewarded for their ethnic identity by being granted permission to express their ethnic origin, which in turn involves a frank reassessment of their culture. In that vein, some professionals have taken on the indigenous struggle for self-recognition in order to gain advantages and achieve the same positions in the socioeconomic hierarchy of those who do not claim any (indigenous) ethnicity. This process impacts their relationships at a social macro-level and is hence not without controversy.

Of course, even while activities are linked to profit from using Mayan culture-specific knowledge, they are quite divisive as to whether or not the idea to implement such activities came from government officials with ethnic origins. In fact, many professional fields (even academia) often comment negatively on those who benefit from their ethnicity. There is even symbolic violence in the form of name-calling that labels Maya professionals "Neo-Maya" or "high-class Indians," whenever indigenous individuals are in positions where they can take advantage of their cultural background or benefit from their ethnicity, such as in any of the ways previously mentioned (in government, business, artisanship, tourism, etc.), or when such individuals try to obtain financing for projects or resources such as scholarships and other aid for minorities. Those activities connected to competition are becoming typical in all workplace environments; especially obtaining funds for projects has become competitive because of the global economic crisis. It must also be said that, in certain cases, some Maya use ethnicity as a strategy to accuse others of disrespect, even when an insult did not really occur or touched ethnolinguistic issues. Such incidents serve to underline that the importance of the "ethnic variable" is not to be underestimated. In any case, it should be recognized that ethnicity and cultural specificity are used as a rational means to an end, one that is clearly designed to grant access to social spaces that historically were denied to the Maya.

Works Cited

Barth, F. (ed.) 1976. *Los grupos étnicos y sus fronteras*. México, D.F.: Fondo de Cultura Económica.
Bertraux, D. 2005. *Le récit de vie*. Paris: Armand Colin.
Bourdieu, P. 1979. *La Distinction: Critique sociale du jugement*. Paris: Editions de Minuit.
Cornelius, W. et al. (eds.) 2007. *Mayan Journeys: US-Bound Migration from a New Sending Community*. La Jolla, CA: University of California at San Diego, Center for Comparative Immigration Studies.

Duarte, A. R. 2008. Imaginando a los mayas de hoy: autorrepresentación y política. *Estudios de cultura maya*, 32. México, D.F.: UNAM, 39–62.

Gaultier, S. 2003. Migración rural y nuevos sistemas localizados de producción: las evoluciones territoriales de la región peninsular, in *Territorios, actores y poder: Regionalismos emergentes en México*, edited by J. Preciado Coronado et al. México, D.F.: UdeG–UADY.

Geertz, C. 1963. The integrative revolution, in *Old Societies and New States*, edited by C. Geertz. New York: Free Press.

González Casanova, P. 1975. *Sociología de la Explotación*. México, D.F.: Siglo Veintiuno.

Gutiérrez Chong, N. 2001. *Mitos Nacionalistas e identidades étnicas: Los intelectuales indígenas y el Estado mexicano*. México, D.F.: FONCA, IIS–UNAM, Plaza and Valdés Editores.

Gutiérrez Chong, N. 2007. Ethnic origins and indigenous people: an approach from Latin America, in *Nationalism and Ethnosymbolism: History, Culture and Ethnicity in the Formation of Nations*, edited by A. S. Leoussi and S. Grosby. Edinburgh: Edinburgh University Press.

INEGI. 2001. *XII censo de población y vivienda 2000*. Aguascalientes, México: Instituto Nacional de Estadística, Geografía e Informática.

INEGI. 2006. *II conteo de población y vivienda 2005*. Aguascalientes, México: Instituto Nacional de Estadística, Geografía e Informática.

Lizama Quijano, J. 2007. *Estar en el mundo: Procesos culturales, estrategias económicas y dinámicas identitarias entre los mayas Yucatecos*. México, D.F.: CIESAS–Miguel Ángel Porrúa.

López Santillán, R. 2006. Pasado rural y pobre, presente de clase media urbana: trayectorias de ascenso social entre mayas yucatecos residentes en Mérida. *Península,* 1(2), 107–28.

López Santillán, R. 2007. Fronteras étnicas, formas de minorización y experiencias de violencia simbólica entre los profesionistas mayas yucatecos residentes en Mérida. *Península,* 2(1), 137–55.

López Santillán, R. 2008. *Clase media en el Distrito Federal: Recomposición de su espacio social y urbano (1970–2000)*. México, D.F.: UACSHUM–UNAM.

Malešević, S. 2004. *The Sociology of Ethnicity*. London, Thousand Oaks, and New Delhi: Sage.

Marín, G. 2010. Turismo, globalización y mercantilización del espacio y la cultura en la Riviera Maya: un acercamiento a tres escenarios, in *Etnia, Lengua y Territorio. El Sureste ante la globalización*, edited by R. López Santillán. México, D.F.: CEPHCIS–UNAM.

Mattiace, S. L. 2007. *We Are Like the Wind: Ethnic Mobilization among the Maya of Yucatán*. Paper to the 27th LASA Conference, Montreal, September 5–9.

Pozas, R. and Horcasitas, I. 1995 [1971]. *Los indios en las clases sociales de México*. México, D.F.: Siglo Veintiuno.

Quintal, E. F. 2005. Wayano'one: Aquí estamos. La fuerza silenciosa de los mayas excluidos, in *Visiones de la diversidad: Relaciones interétnicas e identidades indígenas en el México actual*, vol. 2, edited by M.A. Corral. México, D.F.: INAH: 289–371.

Reséndiz García, R. 2001. Biografía: proceso y nudos teórico-metodológicos, in *Observar, escuchar y comprender: Sobre la tradición cualitativa en la investigación social*, edited by M. L. Tarrés. México, D.F.: FLACSO–COLMEX and Miguel Ángel Porrúa.

Ruz, M. H. (ed.) 2002. *Los mayas peninsulares: un perfil socieconómico*. México, D.F.: UNAM–IIFL.

Serrano Carreto, E., Ambríz Osorio, A. and Fernández Ham, P. (ed.) 2002. *Indicadores socioeconómicos de los pueblos indígenas de México 2002*. México, D.F.: INI–PNUD–CONAPO.

Stavenhagen, R. 1969. *Las clases sociales en las sociedades agrarias*. México, D.F.: Siglo Veintiuno.

Thies, S. and Kaltmeier, O. 2009. From the flap of a butterfly's wing in Brazil to a tornado in Texas? Approaching the field of identity politics and its fractal topography, in *E Pluribus Unum? National and Transnational Identities in the Americas / Identidades nacionales y transnacionales en las Américas*, edited by S. Thies and J. Raab. Münster and Tempe, AZ: Lit Verlag and Bilingual Review Press, 25–46.

Wieviorka, M. 2003. Diferencias culturales, racismo y democracia, in *Políticas de identidades y diferencias sociales en tiempos de globalización*, edited by D. Mato. Caracas: FACES–UCV: 17–32.

Chapter 12

A City of Newcomers: Narratives of Ethnic Diversity in Vancouver

Alicia Menéndez Tarrazo

In recent years, Vancouver has been defined, represented and marketed as a city of newcomers: a diverse, multicultural city in which visitors and immigrants from all backgrounds are welcome. The latest census data reveal that foreign-born people account for 45.6 per cent of the population of the City of Vancouver and 39.6 per cent of the population of the Vancouver metropolitan area (Statistics Canada 2007b: 18–34). Two-thirds of the adult population are linked to immigration in some way, either because they immigrated themselves or because they were born in Canada to immigrant parents (UN-HABITAT 2006). Forty-one per cent of the residents of the Vancouver metropolitan area do not consider either English or French their mother tongue. The most important non-official mother tongues are the Chinese languages (including Mandarin, Cantonese, Hakka, and Taiwanese, among others), followed by Punjabi, Tagalog, Korean, German, Spanish, Farsi, and Hindi (Statistics Canada 2007a: 6–12).

Vancouver's Civic Policy on Multicultural Relations recognizes ethnic and cultural diversity as a strength and a source of enrichment and aims to guarantee freedom from prejudice at all levels of social interaction. Ethnic neighborhoods and communities are encouraged to celebrate their customs, their heritage, and their cultural and religious festivals. The idea of multiculturalism is promoted as an asset, one of the aspects that make Vancouver one of the most livable cities in the world,[1] attractive for tourists and prospective residents alike. Travel guides characterize it as "a modern, multicultural metropolis that showcases tastes, cultures, art and influences from around the world" (Tourism BC 2008), and the City of Vancouver boasts an inclusive hiring policy that seeks to encourage diversity among its employees (City of Vancouver 2008b). As Vancouver has become the site of international events like the 2010 Olympic and Paralympic Winter Games, the marketing of diversity and multiculturalism has reached a global level.

---

1   Since 2005, Vancouver has consistently topped the prestigious "Global Liveability Ranking," developed by *The Economist* Intelligence Unit to assess the quality of life of cities around the world. Other well-known liveability rankings, like Mercer's "Worldwide Quality of Living Survey" and *Monocle* magazine's "Global Quality of Life Survey" often place Vancouver among their top ten or top fifteen cities.

Unpacking the Concept of "Newcomer": Immigration to Vancouver throughout History

Immigration to Canada became especially prominent in the 1970s and 1980s, after the removal of draconian legal restrictions, at a time when the country was taking its first steps towards an official multiculturalism policy. It is probably for this reason that the characterization of Vancouver as a city of newcomers is often embedded in discussions about the adoption of the multicultural model, about the increase in immigration which took place in the last thirty years of the twentieth century, and about the subsequent ethnic diversification of Canadian society. In other words, the notion of "newcomer" tends to be intuitively associated with recent immigrants.

In reality, however, Vancouver has always been a city of newcomers. The text which inspired the title of this chapter, *A Newcomer's Guide to the City of Vancouver*, states that "Vancouver has been a city of newcomers from many different cultures since non-Native peoples began settling in the area" (City of Vancouver 2002: 7). Here, the concept of newcomer is understood as referring to all the non-Native (i.e. non-First Nations) peoples who have ever arrived and settled in Vancouver. This notion puts into a different perspective the oppression and marginalization of ethnic minorities whom the white Anglo-Saxon settlers labeled as "immigrants," "foreigners," "outsiders," and "permanent newcomers" (Fernando 2006: 21–33, McDonald 1996: 201–12) since, at some point in history, white Europeans were newcomers too. Most importantly, newcomers from other areas were *already there* even before the city of Vancouver existed: when Vancouver was incorporated in 1886, some Chinese settlements had already been established in the area (Ng 1999: 10), and the first East Indian immigrants arrived soon thereafter. This means that, since its early years, Vancouver has always shown some degree of ethnic diversity[2]—we could say that Vancouver was a city of newcomers even before it became a city.

It is important to note, however, that not all types of newcomers have always been welcome in Vancouver: historically, non-white immigrants have often been met with suspicion, hostility, and rejection by the white majority. In a study of public attitudes and policies towards Asian people in British Columbia—eloquently titled *White Canada Forever*—W. Peter Ward observes that during the second half of the nineteenth and the first half of the twentieth century, when non-white immigrants came mainly from Asia, anti-Orientalism was endemic, not only in Vancouver, but also in the whole province of British Columbia (2002: 167). The emergence and predominance of this anti-Orientalist feeling has been explained as the direct consequence of economic competition: Chinese and East Indian workers were paid lower wages and blamed for taking jobs away from white laborers. Ward

---

2   Even though the focus here is on newcomers, it should be noted that First Nations peoples have always been present in the city of Vancouver and contributed to its diverse ethnocultural character.

expresses his dissatisfaction with this argument, pointing to the fact that anti-Orientalism did not increase during economic crises or decrease during periods of economic bonanza (2002: xxiii). Instead, he argues that the main cause of anti-Orientalism was "the white community's psychological tendency to the ideal of the homogeneous society" (2002: 54). By 1914, due to massive immigration from the British Isles, Vancouver's population was predominantly white and Anglo-Saxon. This Anglo-Canadian majority "yearned for a racially homogeneous society ... [and] feared that heterogeneity would destroy their capacity to perpetuate their values and traditions, their laws and institutions" (2002: 169). Ward describes the generalized anxiety and prejudice against the non-white population as a "racist consensus" (2002: 167) that was so prevalent as to become the "cultural norm" (2002: 168).

Legislation during this period was designed to make immigration difficult or even impossible. Federal laws like the Chinese Immigration Act of 1885 established serious restrictions on Chinese immigration and imposed an ever-rising head tax on Chinese individuals entering the country: originally a fifty-dollar tax, it was raised to the prohibitive amount of five hundred dollars in 1903 (Fernando 2006: 23, Ng 1999: 11).[3] The Chinese Immigration Act of 1923, often referred to as the Chinese Exclusion Act, effectively stopped Chinese immigration to Canada until its repeal in 1947. East Indians were subject to similar legal restrictions, such as the Continuous Passage Rule of 1908, which required them to arrive in Canada by continuous passage from India at a time when there was no direct steamship service between the two countries (Henry and Tator 1985: 322).[4] At the same time, provincial laws curtailed the rights and liberties of non-white Canadian residents, excluding them from certain types of employment, imposing extra taxes on their income, and denying them the right to vote (Fernando 2006: 23).

Members of ethnic minorities were the object of all kinds of discriminatory attitudes and behaviors, sometimes even violence, as in the Vancouver race riots of 1907, when a white mob tore through Chinatown and Japantown, terrorizing their inhabitants and destroying their property. In 1942 Japanese immigrants who had settled along the Pacific coast were accused of spying for the Japanese army, labeled as enemy aliens, ordered to evacuate their homes, and sent to work camps in the interior of the country. Their property and personal effects were confiscated, families were separated, the Japanese Canadian community was destroyed, and some of them were even deported to Japan. It was, numerous critics agree, the

---

3   In 2006, after a long struggle by the Chinese Canadian community, Prime Minister Stephen Harper offered a formal apology for the use of the head tax and for the exclusion of Chinese immigrants. Symbolic compensation payments were awarded to surviving tax payers and to spouses of deceased payers.

4   A direct consequence of this policy was the Komagata Maru incident, in which 376 men of East Indian origin unsuccessfully attempted to enter Canada in a steamship chartered from Hong Kong. Not being allowed to dock in Vancouver, the ship was eventually forced to turn back.

most flagrant and shameful expression of racism in the history of Canada (Adachi 1976, Henry and Tator 1985: 322, Miki 2004, Ward 2002: xii).[5]

After the Second World War the situation of non-white Canadians began to improve slightly, at least at the level of legislation. In 1947 the franchise was extended to Asian Canadians, and the Chinese Exclusion Act was repealed, although some restrictions were in place until 1967, when the racist provisions that sought to limit or prevent the immigration of non-white individuals were removed from the Canadian Immigration Act. This measure encouraged renewed immigration from traditional places like India and China, at the same time attracting newcomers from other, non-traditional areas in Asia, Africa, and the Caribbean (Fernando 2006: 27, Henry and Tator 1985: 323). In the 1960s Canada was becoming increasingly diverse, yet it was still conceived as a bicultural society, a country whose two founding cultures were the British and the French. It was only in the 1970s and 1980s that Canada began to consider itself a multicultural country; the notion of multiculturalism was written into the legislation for the first time in 1971.

Multiculturalism: Official Policies and Unofficial Discourses

The Canadian Multiculturalism Act, whereby Canada defined itself as a multicultural, multiethnic, and multiracial society, was passed in 1985. The Act recognized the important role of multiculturalism in the construction of Canadian identity and of Canada's future. It guaranteed respect towards different cultures and vowed to promote their development and continuity, ensuring the full participation of all individuals and communities in Canadian society regardless of their origin (Department of Justice 1985).

In recent years, the Multiculturalism Act has become an object of debate: its critics believe that Canada's complacency and celebratory attitude to its reinvention as a multicultural country disregards a racist past that needs to be acknowledged before it can be left behind. Shanti Fernando observes that "aspects of Canadian history that reflect a more openly racist time have been ignored in order to emphasize its current status as a multicultural country that supports diversity" (2006: 21) and argues that "the ability to change racist laws was a positive step, but the denial that any such laws existed is dangerous as it means the legacy of a colonial past has not been addressed" (2006: 25). Sheila Van Wyck and Ian Donaldson express a similar concern:

> Despite Canada's *international reputation and self-image* as the world's most peaceable multicultural liberal democracy—a *national myth* that arguably began

---

5   Redress for the unjust treatment of Japanese Canadians was achieved in 1988. Prime Minister Brian Mulroney apologized to the Japanese Canadian community, and monetary compensations were awarded to the individuals and communities affected by the evacuation.

to take root in the public imagination after the 1969 official languages policy and the 1971 multiculturalism policy—history tells a more complex and often *painful* story. Interestingly, Canadians tend to suffer from a kind of *collective amnesia* with respect to such history. (2006: 141, emphasis added)

Vancouver's Civic Policy on Multicultural Relations was approved in 1988 in a similar spirit to that of the Canadian Multiculturalism Act. Diversity was recognized as one of Vancouver's main strengths and a source of enrichment for its residents. The access to civic services was guaranteed to all citizens, independently of their country of origin, their ethnicity, the language they spoke, or their difficulties to communicate in English (City of Vancouver 2006b). Outreach programs were put in place to assist visible minorities, immigrants, and refugees, facilitate the integration of newcomers, remove barriers to their participation in civic processes, and foster community formation. Vancouver's diverse neighborhoods and communities were encouraged to preserve their heritage and customs and to celebrate their cultural and religious festivals (City of Vancouver 2006a).

These legislative measures and public policies constitute what we could call an *official* narrative of ethnic diversity in Vancouver. Official discourse emanates from those institutions invested with the power to construct, circulate, and enforce a dominant or hegemonic narrative of the city. Legislation is, by definition, an important part of that discourse, but so are other, non-prescriptive pieces of official literature, such as the promotional texts produced by government agencies in order to attract visitors, immigrants, prospective employees, and foreign investors.[6] In contrast to this official discourse produced and circulated by public institutions and organisms, there is a different type of discourse that we could call *unofficial:* the individual and collective micronarratives of the city created by its inhabitants, by the ways in which urban dwellers use, perceive, represent, conceptualize, explain, or imagine urban spaces in their everyday lives.

Urban Imaginaries: Contrasting Perceptions of Ethnic Diversity

Between the official and the unofficial lies another very important narrative that is sometimes difficult to categorize: the collective urban imaginary, a set of widespread perceptions, assumptions and generalizations about the city which may or may not be sanctioned by official or institutional sources, but which nevertheless constitute a very important part of the reality of the city. Several scholars have noted that urban imaginaries are didactic, prescriptive, and normative: they determine our perception of the urban reality and our way of inhabiting the city (Bounds 2004: 122–23, Donald 2000: 50; Shields 1996: 229). Furthermore, these imaginaries are constructed discursively: they emerge from the intersection of knowledge and

---

6   For specific examples related to the city of Vancouver, see City of Vancouver 2008b; Government of Canada 2010; Tourism BC 2008.

power, which means that they often serve specific interests or ideologies (Bridge and Watson 2000: 16, Shields 1996: 227). Above I have discussed an example of how this powerful imaginary construction may permeate the official discourse and determine political decisions with far-reaching consequences: at some points in the history of Vancouver, when the dominant perception among the ethnic majority was that there were too many Asian immigrants threatening white Anglo-Saxon predominance, legislation was passed to stop what was perceived as an "Oriental invasion." Vancouver liked to imagine itself as a homogeneous, white, Anglo-Saxon settler city, and the legislators worked to make this imaginary real.

Nowadays, the situation is considerably more complex. The City of Vancouver's official narrative regarding ethnic diversity is one of proud multiculturalism and harmonious multiethnic coexistence, as evidenced in its legislation and official literature. Popular attitudes towards diversity, however, are varied and contradictory. On the one hand, Vancouver's urban imaginary is dominated by a generalized perception of diversity as something positive, desirable and beneficial for the city and its inhabitants. This narrative goes hand in hand with the official discourse of civic policy and is perpetuated by the real estate, tourism, and service sectors: realtors, restaurant owners, shopping districts, and heritage areas all promote the idea of diversity as a way of attracting customers. As diversity becomes a marketable commodity, the discourse of multiculturalism often becomes one of glamorous cosmopolitanism and sophistication—an example of *Selling EthniCity* at its best. On the other hand, at the same time that diversity is celebrated as positive, some of Vancouver's most ethnically diverse neighborhoods are constructed as problematic and undesirable. East Vancouver, for example, has been codified in the urban imaginary as a no-go area because of socioeconomic factors such as poverty, incidences of crime and drug addiction, unemployment, and, paradoxically, a predominance of immigrant population. It is true that this area faces all these problems, but it is not because of the immigrant population. In fact, the opposite seems to be true: it is arguably the socioeconomic problems in this area that make it the last affordable place to live, thus attracting new immigrants and working-class families who cannot afford property prices elsewhere. This points towards the possibility that the stigmatization of this neighborhood in the collective imagination might emerge from the intersection of ethnicity and social class. In *After the Cosmopolitan? Multicultural Cities and the Future of Racism* (2005), Michael Keith argues that the notion of multiculturalism may have positive or negative connotations depending on the context: "the street, the ghetto, the *banlieue* and the cultural quarter make visible very different sorts of urbanism and very different sorts of multiculturalism" (2005: 10). These seem to be the dynamics involved in the imaginary construction of East Vancouver as a dangerous area: in middle-class neighborhoods, multiculturalism is perceived as something to be consumed, safe, and manageable; in working-class or lower-income areas, however, diversity acquires a negative connotation, as ethnic minorities are cast as Others and perceived as an anomaly or even a threat to peaceful multicultural coexistence.

Literary Perspectives

It is common for contemporary literary representations of Vancouver to challenge most of the dominant narratives and preconceived notions about the city. The excerpts I am going to analyze here have been taken from literary texts written in the last thirty years, most of them after the Canadian Multiculturalism Act was passed, and they offer very interesting perspectives on relevant issues like ethnocultural diversity, exclusion and inclusion, racism, prejudice, and social conflict. I have divided the extracts into three categories: firstly, I am going to examine four contemporary novels set in the times before the Multiculturalism Act: *Obasan* (1981) by Joy Kogawa, *Making a Stone of the Heart* (2002) by Cynthia Flood, *Disappearing Moon Cafe* (1990) by Sky Lee, and *All that Matters* (2004) by Wayson Choy. I will look at the different ways in which these literary texts respond to the dominant narrative of that time: the ideal of the white, Anglo-Saxon, settler society. Secondly, I am going to discuss two texts set in the 1980s and 1990s, Carmen Rodríguez's "Breaking the Ice" (1997) and Anita Rau Badami's *Can You Hear the Nightbird Call?* (2006), focusing on the contrast between the multiculturalism of the official policy and that of everyday reality. Finally, I will comment on three texts set in present-day East Vancouver that question the symbolic construction of this neighborhood as a dangerous, sordid area: *Blood Sports* (2006) by Eden Robinson, *The End of East* (2007) by Jen Sookfong Lee, and "Raise Shit" (1999) by Bud Osborn.

Joy Kogawa was the first novelist to choose the Japanese evacuation of 1942 as her subject matter. Kerri Sakamoto notes in her introduction to recent editions of the novel that, before *Obasan* was published in 1981, the plight of the Japanese Canadians during and after the Second World War did not even appear in history textbooks (2006: viii). Kogawa gives a voice to the shattered Japanese Canadian community and foregrounds the entrenchment of racism in the Canadian society and institutions of the time. Shortly after the Pearl Harbor bombing, one of her protagonists writes in her diary: "Thank God we live in a democracy and not under an officially racist regime" (Kogawa 2006: 76). This comment, full of proleptic irony, becomes a powerful statement against the cruel, undemocratic mistreatment of Japanese Canadians that began soon thereafter. *Obasan* describes all the events leading up to the evacuation in great detail, calling attention to their human side, to their emotional and psychological effects on the population, and to the real motives behind them:

> Things are changing so fast. First, all the Japanese men—the ones who were born in Japan and haven't been able to get their citizenship yet—are being rounded up, one hundred or so at a time. ... It isn't just a matter of fear of sabotage or military necessity any more, *it's outright race persecution.* ... And the horrors that some of the young girls are facing—outraged by men in uniform. You wouldn't believe it, Nesan. You have to be right here in the middle of it to really

know. The men are afraid to go and leave their wives behind. (2006: 79–81, emphasis added)

The novel also reveals the contradictions inherent in the official discourse on people of Asian origin:

One letter in the papers says that in order to preserve the "British way of life," they should send us all away. We're a "lower order of people." In one breath we are damned for being "inassimilable" and the next there's fear that we'll assimilate. One reporter points to those among us who are living in poverty and says "No British subject would live in such conditions." Then if we improve our lot, another says "There is danger that they will enter our better neighbourhoods." (2006: 81)

The racist stereotypes at work in this passage have been documented outside fiction by Robert A.J. McDonald and Peter Ward, who observe that the construction of the Asian immigrants as inassimilable turned them into a threat to the white majority because of their capacity to "inundate British Columbia and destroy its collective character as a land of White, European-based settlers" (McDonald 1996: 206). Assimilation, on the other hand, was deemed to be impossible: the cultural differences between whites and non-whites were perceived as insurmountable, and the possibility of assimilation through miscegenation was "so remote to many whites that it scarcely entered their minds" (Ward 2002: 13).

Making a Stone of the Heart (2002) tells the intertwined stories of two characters whose lives span the whole twentieth century; in the background, the small mill town of Vancouver grows into the city it is today. Cynthia Flood illustrates the ethnic tensions between the white majority and the non-white minorities and highlights the presence, throughout Vancouver's history, of the oft-forgotten First Nations people. In 1912, for example, a schoolteacher tells one of her students: "You live in a *city*, Owen Jones, not a wilderness. … And the savages gone, thank goodness, with their drink and their lice and idleness. Vancouver's no place for beasts now" (2002: 127, emphasis in the original). The child, however, knows that there is a whole Squamish village under Burrard Bridge (2002: 133), and thus the homogeneous white city described by the teacher is revealed as a product of the white majority's imagination.

Later on in the novel, the racial tensions brought about by the Second World War, related by Joy Kogawa in *Obasan,* are described from the perspective of a white Anglo-Saxon woman who finds employment with a German tailor:

This tall spare woman said at once, "I am German. If you do not wish, let us not waste our time. … That is Miss [Jenny] Ko. … Jenny's mother is in the camp at Slocan. She herself hopes not to be taken. Her father—he is dead but was not Japanese, you see. …" At lunch Dora ate her baloney sandwiches as Jenny tweaked a salted plum or a riceball from a lacquered box, while the tailor and her daughter munched dark rye layered with strong-smelling cheese. Soon, like the others, Dora drank tea from tiny flowered cups. Setting her treasures gently back on the shelf, Jenny said, "One less thing for them to smash, if someone tells and they search my room." (2002: 200–201)

Despite the pervasiveness of racial prejudice and racist behaviors, these working-class women manage to create a diverse space of peaceful multiethnic coexistence which, we may assume, was probably not unique in the Vancouver of the time. This excerpt challenges the profiling of German and Japanese Canadians as enemy aliens, and belies the generalized assumption that different ethnic groups inhabited completely separate spheres of social space.

Chinese Canadian authors have also paid attention to the mistreatment of their own and other non-white communities in wartime and postwar Vancouver. In Sky Lee's *Disappearing Moon Cafe* (1990), a family saga which explores the immigrant experience across different generations, a young Chinese woman becomes "the object of much scorn and derision. East Vancouver wartime youths not being the most big-hearted and open-minded about the cultural diversity they called 'chinks, japs, wops and hindoos!'" (1990: 175). Similarly, in *All that Matters* (2004), Wayson Choy's young narrator explains that "Indians and blacks, Asians of every variety—all those who were not permitted entry into regular hospitals—ended up, if there were any spaces available, at the segregated Home for Chinese, or at St. Joseph's Oriental Hospital, or in the grey-painted basement of St. Paul's" (2004: 362).

These historical novels portray a Vancouver which was definitely not as homogeneous as the white majority liked to imagine it: ethnic diversity existed in the city since its inception. The dominant narrative that denied this diversity also contributed to the oppression and marginalization of non-white ethnic minorities. The extracts cited above reveal the contradictions between the discourse of the white settler society and the multiethnic diversity of everyday urban life.

Literary works set in contemporary Vancouver display a similar tendency to challenge dominant narratives about the city, like the celebration of multiethnic diversity as an integral part of its identity. In Carmen Rodríguez's short story "Breaking the Ice", set in the early 1990s, the official multiculturalism policy does not eradicate racist ideas and attitudes at the level of everyday social interaction, as evidenced by the following exchange between three women of Italian, Portuguese, and Chilean origin:

> I don't know what to do, Rosa said; it turns out that Luzia has been misbehaving, now she's come up with a boyfriend, a Negro boy. … Can you imagine? Black as coal. Preto, preto. Oh, meu Deus, what have I done to deserve this?
> I don't know, Silvia, Signora Carmela says. I don't know what to think. I had never seen a black person until I came to Vancouver, and it's only in the last few years that you see more of them. … I've never talked to anybody who's black. … Rosa hasn't either. … That stupid Luzia! Why a black boy when there are so many Canadian and Portuguese boys? (1997: 112–14)

The above passage is especially interesting if we consider the fact that most Canadians of Anglo-Saxon descent would probably perceive its protagonists as non-white, visible minority individuals, and maybe even treat them with the same derision they direct at the black boy. In this story, ethnic minorities, no matter how similar in their marginalization by the white majority, are also guilty of racial prejudice against each other.

Even more thought-provoking is the following extract from Anita Rau Badami's 2006 novel *Can You Hear the Nightbird Call?*, in which people of Indian origin are rejected and marginalized by people of their own ethnicity:

> Not only was this world divided, with the desis ranged on one side against the goras, the Asians, the Africans, the Native Americans, but there was a further distinction to be made. There were the *real Indians*, like the Bhats and the Patels and the Majumdars, those travellers who had come *from Bombay or Calcutta or Aurangabad*. And then there were *others*, like JB, who had arrived in Canada *from East Africa*, traumatized by Idi Amin's decision to confiscate their property and belongings and eject them from Uganda with nothing more than the clothes on their backs. According to Leela *their two-hundred-year stopover in East Africa disqualified them from being called Indian*: they had forfeited the right to return. (2006: 262, emphasis added)

The dynamics of exclusion at play here point to the notion of diasporic identity, defined by Avtar Brah as a transnational network of identification with "imagined" and "encountered" communities (1996: 192), as well as to its limitations. Ien Ang, in her discussion of Chinese diasporic identity, argues that an excessive symbolic orientation toward the "homeland" is problematic, because it presents the country of origin as the cultural epicenter of the diaspora, the ideal against which diasporic identities are measured in terms of cultural purity and authenticity (2001: 32). In the passage cited above, this privileging of the homeland results in Indian immigrants from India being considered purer, more authentic, and *more Indian* than those who occupy more distant positions in the spectrum of the Indian diaspora.

These are only two examples of the way contemporary Vancouver writers call into question the ideal of the peacefully multicultural society. In these texts and others, racial prejudice is no longer overt or officially sanctioned, but it still informs the social behavior of many Vancouver residents. As Shanti Fernando puts it, "even though Canadians seemed to accept multiculturalism at face value, the presence of many cultures was a cause of tension" (2006: 28).

The last group of texts I am going to look into comprises three very different literary works that coincide in challenging the imaginary construction of present-day East Vancouver as a dangerous, undesirable neighborhood. Eden Robinson's *Blood Sports* (2006), for example, is an extremely harsh novel about crime, poverty, violence, and drug addiction; yet the author refuses to let these motifs dominate her representation of East Vancouver. The neighborhood and its inhabitants are described in the following way:

Paulie had ambitions to get them a house in a "nice" neighbourhood, but Tom loved East Vancouver. He'd grown up here, and yeah, it was a little rough around the edges, but they weren't exactly pruning roses and taking high tea. ... Because of its slightly seedy reputation, East Van was one of the last affordable places to live in Vancouver, so it attracted a mix of anarchists and activists, blue-collar families and immigrants. The hippies who couldn't afford Kitsilano had also migrated east, bringing organic co-ops and hemp shops into the mix. ... You could find a Jamaican jerk shack beside an Ethiopian vegetarian café beside a hydroponics bong palace. (2006: 31–4)

This excerpt is a good representative sample of the general tone of the novel, and also of the approach most authors adopt when writing about East Vancouver: acknowledging its problems while at the same time drawing attention to its more positive aspects, such as a strong sense of community, intense political involvement and activism, solidarity, and the acceptance of very different cultures and lifestyles.

In *The End of East* (2007), Jen Sookfong Lee looks very closely at East Vancouver, her own neighborhood, focusing on its ugliest areas but always finding the beauty hidden behind them:

It is really the East Side that is Vancouver for me—the netless basketball hoops stranded in their concrete courts, the stained stucco on the sides of squat apartment buildings, the spit-crossed sidewalks that seem to lead everywhere you'd ever want to go until you realize that they can only end in ocean or mountain or trees. ... I walk this city every day, sidestep the garbage, hold my breath through the alleys. But even in the dirtiest of places, where the sidewalk is covered with gum and the hum of traffic and city noise is so loud that you can't even hear your own footsteps, you can always look north and see the mountains. (2007: 191–92)

Bud Osborn is another East Vancouver author who strives to shed a more positive light on this area through his writing and social activism. His poetry deals with the situation in the Downtown Eastside and the problems encountered by its inhabitants in their everyday lives. The Downtown Eastside is the most deprived neighbourhood in Vancouver and the poorest postal code in Canada: it is afflicted with endemic unemployment, lack of housing and homelessness, and high levels of drug addiction and mental illness (City of Vancouver 2008a). Seventy percent of the population suffer from hepatitis, and HIV infection levels reach 30 percent—the highest rate in the so-called First World, similar to that of African countries like Botswana (Pivot Legal Society 2008). Osborn's poetry is dark but hopeful, always focusing on the human side of a neighborhood whose inhabitants are often dehumanized through their representation in the mass media and political discourse. In poems like "Raise shit" (1999), aptly subtitled "Downtown Eastside poem of resistance", he reclaims the dignity of the Downtown Eastside while calling its population to protest their marginalization:

we have become a community of prophets in the downtown eastside
rebuking the system
and speaking hope and possibility into situations
of apparent impossibility

a first nations' man recently told me
he had come to the downtown eastside to die
he heard the propaganda
that this is only a place of death disease and despair
and since his life had become a hopeless misery
he came here specifically to die
but he said
since living in the downtown eastside
what with the people he has met
and the groups he has found
he now wants very much to live
and his words go directly
to the heart of what makes for real community
...
we resist
because we are a community
...
and it is to our credit that this is so
for it is from our
prophetic courageous conflictual and loving
unity
that our community
raises shit
and resists (1999: 27–36)

The poets and novelists who choose to write about East Vancouver not only undermine preconceived notions about this area but, most importantly, they also create and circulate alternative, positive representations that counteract the negative images deeply rooted in the collective imagination of the city. Robinson, Sookfong Lee, and Osborn call into question the stigmatization of East Vancouver in much the same way that Kogawa, Flood, Lee, and Choy questioned the past construction of Vancouver as a homogeneous white city, and Rodríguez and Badami challenged its present construction as the poster city for Canadian multiculturalism.

Conclusion

This chapter constitutes an overview of some of the most relevant narratives of Vancouver's ethnic diversity. I have emphasized the ways in which different narratives support, question, or contradict each other because I firmly believe one of the most important aspects of urban ethnic imaginaries is that they can and must be challenged.

Official narratives are easy to access, through legislation and institutional literature. Unofficial narratives, be they individual or collective, hegemonic or marginal, are harder to isolate and analyze. This is why literary representations of the city become invaluable resources for the transdisciplinary field of Urban Studies. Literary texts document the individual or collective urban experience as no other text does, allowing us to identify the contradictions and discontinuities between the official discourse and the everyday reality of the city.

Contemporary Vancouver authors are engaged in a continuous process of rewriting, reinterpreting, and reinventing the city. Writers of historical fiction like Joy Kogawa, Cynthia Flood, Sky Lee, Wayson Choy, and others have rewritten the history of Vancouver from a contemporary, more inclusive point of view. Their novels reveal the presence of ethnic minorities as well as their marginalization and also their contribution to the development of a city that for a long time denied them citizenship. Thus, literature recovers events that have been ignored and voices that have been silenced by official narratives, becoming an effective antidote to historical amnesia.

Carmen Rodríguez and Anita Rau Badami are just two examples of the many authors whose writing challenges the contemporary narrative of multiethnic bliss. In their works of fiction, the everyday experiences of the individuals and communities who have to deal with racial prejudice and social rejection undermine the uncritical celebration of Vancouver's cultural diversity. Conversely, Eden Robinson, Jen Sookfong Lee, Bud Osborn, and others celebrate in their writing the positive qualities of a neighbourhood traditionally burdened with social stigma, proving that the work of novelists and poets does not always involve revealing the dark underside of dominant narratives.

Ultimately, and perhaps most importantly, literary narratives question the issue of authority and undermine the opposition between urban reality and urban representation: in the creative tension between different ways of imagining the city, urban space reveals itself as a discursive construction whose symbolic component cannot be ignored.

Works Cited

Adachi, K. 1976. *The Enemy that Never Was: A History of the Japanese Canadians*. Toronto: McClelland and Stewart.

Ang, I. 2001. *On Not Speaking Chinese: Living between Asia and the West*. London: Routledge.

Badami, A. R. 2006. *Can You Hear the Nightbird Call?* Toronto: Vintage Canada.

Bounds, M. 2004. *Urban Social Theory: City, Self and Society*. Oxford: Oxford University Press.

Brah, A. 1996. *Cartographies of Diaspora: Contesting Identities*. Florence: Routledge.

Bridge, G. and Watson, S. 2006. City imaginaries, in *A Companion to the City*, edited by G. Bridge and S. Watson. Malden, MA: Blackwell, 7–17.

Choy, W. 2004. *All That Matters*. Toronto: Anchor Canada.

City of Vancouver. 2002. *A Newcomer's Guide to the City of Vancouver*. Vancouver.

City of Vancouver. 2006a. *City's Response to Multicultural and Diverse Issue* [Online: City of Vancouver. Community Services]. Available at: http:// vancouver.ca/ COMMSVCS/socialplanning/initiatives/multicult/cityresponse. htm [accessed: July 3, 2010].

City of Vancouver. 2006b. *Civic Policy on Multicultural Relations* [Online: City of Vancouver. Community Services]. Available at: http://vancouver.ca/ COMMSVCS/socialplanning/initiatives/multicult/civicpolicy.htm [accessed: July 1, 2010].

City of Vancouver. 2008a. *Downtown Eastside Revitalization* [Online: City of Vancouver. Community Services]. Available at: http://vancouver.ca/commsvcs/ planning/dtes/ index.htm. [accessed: July 3, 2010].

City of Vancouver. 2008b. *Employment: Vancouver!* [Online: City of Vancouver. Employment]. Available at: http://vancouver.ca/humanresources/index.htm [accessed: July 3, 2010].

Department of Justice. 1985. *Canadian Multiculturalism Act* [Online: Department of Justice Canada]. Available at: http://laws-lois.justice.gc.ca/en/C-18.7/ FullText.html [accessed: July 5, 2010].

Donald, J. 2000. The immaterial city: representation, imagination and media technologies, in *A Companion to the City*, edited by G. Bridge and S. Watson. Malden, MA: Blackwell, 46–54.

Flood, C. 2002. *Making a Stone of the Heart*. Toronto: Key Porter Books.

Fernando, S. 2006. *Race and the City: Chinese Canadian and Chinese American Political Mobilization*. Vancouver: University of British Columbia Press.

Government of Canada. 2010. *Vancouver: Invest in Canada*. [Online: Government of Canada]. Available at: http://investincanada.gc.ca/eng/explore-our-regions/ western-canada/british-columbia/vancouver.aspx [accessed: July 5, 2010].

Henry, F. and Tator, C. 1985. Racism in Canada: social myths and strategies for change, in *Ethnicity and Ethnic Relations in Canada*, edited by R. M. Bienvenue and J. E. Goldstein. Toronto: Butterworth, 321–35.

Keith, M. 2005. *After the Cosmopolitan? Multicultural Cities and the Future of Racism*. London: Routledge.

Kogawa, J. 1981. *Obasan*. Toronto: Penguin.

Lee, J. S. 2007. *The End of East*. Toronto: Knopf.

Lee, S. 1990. *Disappearing Moon Cafe*. Vancouver: Douglas and McIntyre.

McDonald, R. A. J. 1996. *Making Vancouver: Class, Status and Social Boundaries 1863–1913*. Vancouver: University of British Columbia Press.

Miki, R. 2004. *Redress: Inside the Japanese Canadian Call for Justice*. Vancouver: Raincoast Books.

Ng, W. C. 1999. *The Chinese in Vancouver, 1945–80: The Pursuit of Identity and Power*. Vancouver: University of British Columbia Press.

Osborn, B. 1999. Raise shit, in *Keys to Kingdoms*. Vancouver: Get to the Point, 24–36.

Pivot Legal Society. 2008. *Pivot Legal Society and Vancouver's Downtown Eastside*. [Online]. Available at:http://www.pivotlegal.org/dtes.htm [accessed: July 5, 2010].

Robinson, E. 2006. *Blood Sports*. Toronto: Emblem Editions.

Rodríguez, C. 1997. Breaking the ice, in *And a Body to Remember With*. Vancouver: Arsenal Pulp Press, 109–18.

Sakamoto, K. 2006. Introduction, in *Obasan*, by J. Kogawa. Toronto: Penguin, vii–ix.

Shields, R. 1996. A guide to urban representation and what to do about it: alternative traditions of urban theory, in *Re-Presenting the City: Ethnicity, Capital and Culture in the Twenty-First Century Metropolis*, edited by A. D. King. London: Macmillan Press, 227–52.

Statistics Canada. 2007a. *The Evolving Linguistic Portrait, 2006 Census. 97-555-XIE*. [Online: Statistics Canada: Canada's national statistical agency]. Available at: http://www12.statcan.ca/census-recensement/2006/as-sa/97-555/pdf/97-555-XIE2006001-eng.pdf [accessed: July 2, 2010].

Statistics Canada. 2007b. *Immigration in Canada: A Portrait of the Foreign-born Population, 2006 Census. 97-557-XIE*. [Online: Statistics Canada: Canada's national statistical agency]. Available at: http://www12.statcan.ca/census-recensement/2006/as-sa/97-557/pdf/97-557-XIE2006001.pdf [accessed: July 2, 2010].

Tourism BC. 2008. *Vancouver*. [Online: Super, Natural British Columbia]. Available at: http://www.hellobc.com/en-CA/RegionsCities/Vancouver.htm [accessed: July 2, 2010].

UN-HABITAT. 2006. Vancouver: the world's most liveable city combines multiculturalism with environmental sustainability, in *State of the World's Cities 2006/07*. [Online]. Available at: http://www.unhabitat.org/mediacentre/documents/sowcr2006/SOWCR%2014.pdf [accessed: July 4, 2010].

Van Wyck, S. and Donaldson, I. 2006. Challenges to diversity: a Canadian perspective. *Canadian Diversity / Diversité Canadienne*, 5(1), 140–43.

Ward, W. P. 2002. *White Canada Forever: Popular Attitudes and Public Policy Toward Orientals in British Columbia*. Montreal and Kingston: McGill–Queen's University Press.

PART IV
Gentrification and the Politics of Authenticity

Introduction to Part IV
Gentrification and the Politics of Authenticity

Olaf Kaltmeier

The increasing importance of cultural politics for cities, which is accompanied by a revaluation of downtown areas and the development of centrally located residential neighborhoods for the (often white) upper middle class, massively impacts the sociospatial stratification as well as the social and cultural participation of lower class sectors, which are often ethnically marked in the Americas. After the so-called "white flight" of the white middle class into the suburbs, downtown city centers became subject to processes of ethnicization and pauperization. However, depending on the economic situation, processes of inner-city gentrification have also increased since the 1980s. First formulated in the United States in the 1960s, the concept of gentrification refers to the displacement of less affluent population groups by those enjoying a higher social status—the "gentry"—during the development over time of centrally located urban neighborhoods.

Neo-Marxist approaches explained gentrification processes primarily with references to two intersecting dynamics. On the supply side, return speculations of the real estate-holding capital stimulates these processes (Smith 1987). The consumption-side theory of urban gentrification, on the other hand, underlines the socio-cultural qualities and motives of the gentrifiers, who aim at a comfortable life in city centers equipped with art galleries, delicatessen stores, cafés and restaurants, as well as upscale living opportunities (Zukin 1987, 1989).

The shared connection to cultural production and upscale consumption experiences in the city establishes a tight link between gentrification and postindustrial consumer societies. In this context, artists and cultural producers play an ambivalent role. While literary theory ascribes to them a function of seismograph and catalyst with regard to processes of social change (Zapf 2002), they hold the same function with regard to the material appropriation of space, because cultural producers are often first-stage gentrifiers who take on the role of trend setters in urban revitalization processes. At the same time, they often act as public voices for the concerns of marginalized groups, as illustrated by Martha Rosler's art project "If you lived here ..." (1989), about homelessness and gentrification—alluding to former New York Mayor Ed Koch's phrase "If you can't afford to live here, move!"

All regional and local specificities notwithstanding, gentrification processes are part of a global dynamic inasmuch as cities form nodes in a worldwide net of flows of capital, people, and information. In the context of a global "economy of signs" (Lash and Urry 1994), individual cities attempt to stand their ground within the worldwide net of cities by presenting themselves as "brands." This causes massive identity political conflicts with—often subaltern—lower-class inhabitants. Thus, gentrification not only marks a process of economic displacement, but it is also connected to acts of violence, police actions, as well as mechanisms of discipline and control. In Lima, for example, violent militarized displacements or enforced resettlements of so-called "dangerous classes" (Foucault) accompanied the revitalization of the historic city center. In New Orleans, Hurricane Katrina's devastation of African American lower-class neighborhoods functioned as slum clearance. Moreover, in the revitalized, re-dedicated, and gentrified urban districts culture is often employed as a resource for behavioral regulation—"domestication by cappuccino," as put in a nutshell by Sharon Zukin (1995).

Often, gentrification processes can be realized only partially, as even hegemonic positions must resort to actors of everyday life and their ideas, norms, values, and practices. First, they depend on the generation of a certain public consensus in order to implement their vision of a city and the interventions connected to it. In planning processes, they often have to recur to residents' participation techniques. Dealings with ethnic groups especially involve a considerable potential for identity political conflicts. Secondly, these processes must integrate popular cultural practices and concepts in order to generate the authenticity needed for building the public image of and local identities in the city.

Even though gentrification is connected to "whitening" processes in many cities in the Americas, these developments run parallel to dynamics of migrants moving into the cities, especially into the economically and culturally dynamic urban nodes within the worldwide net of information streams. Accordingly, Saskia Sassen argues that it is not only the transmigration of capital that takes place in this global grid, but also that of people, both rich (i.e., the new transnational professional workforce) and poor (i.e., most migrant workers) and it is a space for the transmigration of cultural forms, for the reterritorialization of "local" subcultures." (1998: 241) Thereby the "creative" and "managerial classes" depend highly on the availability of inexpensive labor for cleaning, childcare, and other services.

This shared global logic notwithstanding, the social stratification of ethnicity in urban areas is subject to distinct regional dynamics. Owing to its central role as global financial center, New York is currently witnessing an expansion of its Chinatown district, as financially potent, transnational young urban professionals from Hong Kong, Taiwan, and Mainland China move into this area, whereas its Latino population is typically employed in the lower-scale service sector (Smith 1996: 286). Miami, on the contrary, has developed into the unofficial capital of Latino cultural production in the United States (Yúdice 2003), generating a Latino cultural elite that stands opposite an equally Latino-dominated class of untrained workers.

One can generally observe that sociospatial polarizations and processes of inclusion and exclusion occur in increasingly narrow spaces and align with the contradictions of the new global division of labor. "Urban ethnic spaces" can thus also take the form of ethnic ghettoes that bear witness to an "urbanization of poverty" and which are disconnected from global streams despite their geographical proximity to the centers. The spatially proximate coexistence of upper-class gated communities and *favelas* in (Brazilian) metropolises provides an extreme example for this phenomenon. Municipalities, in particular in the metropolises, are no longer in the position to ensure social cohesion, meet basic needs, and create equality of opportunities, which inevitably leads to the disappointment of the "metropolitan expectations" of many migrants.

Against the background of the given material infrastructures and specific qualities of these highly segregating urban systems sketched here, one can observe actor-specific appropriation practices. The residents of *favelas*, *barrios populares*, and slums, for instance, develop their proper logic of interaction and practices of appropriating urban space, which reach from substitute economies via the establishment of a "barrio *logos*" (Villa 2000) to cultural political strategies such as graffiti and Hip Hop. These processes bring about new identity formations that Michel Maffesoli (1996) has labeled "urban tribes". However, especially in contrast to the role of the African American ghettoes of the 1960s as locus of a black resistance culture, community formation among these "urban tribes" is to be understood as a highly precarious phenomenon. On the one hand, processes of atomization and disintegration—summarized by Loïc Wacquant as "hyper-ghettoes" (2007)—stand in the way of attempts at community formation, which are often based on territoriality and face-to-face relations in the neighborhood as well as on the binding influence of ethnicity. On the other hand, attempts at self-representation and appropriation of urban spaces are, in turn, subject to commodification processes. Graffiti has arrived in art galleries as a distinct art form, the commercial music industry markets Hip Hop, and even postcolonial literature is subjected to market forces (Huggan 2001). The artist Kalissa Chrish shows how far the boundaries of the marketability of ethnicity in the city can stretch by having turned Wacquant's critical term of the hyper-ghetto into an ironic trademark that specializes in selling autoethnic bling craftwork from the ghettoes and distributes its revenues to community organizations in the "hoods" (Chrish 2010).

Focusing on several recently gentrifying neighborhoods in the USA, the chapter of Eric C. Erbacher takes a critical look at the various, and often problematic, conceptions of ethnicity and ethnic culture used to promote and distinguish a respective area. Thereby Erbacher analyzes the way the media (re)construct safe versions of the ethnic past(s) and present(s) of neighborhoods undergoing gentrification for a seemingly rooted and unique image and identity. Further, this chapter examines various strategies employed by both gentrifiers and institutional actors to stabilize and commodify these new neighborhood identities with consumable experiences such as ethnic festivals and ethnic foods.

Taking the example of the Mexican metropolis of Guadalajara, the sociologist Ulises Zarazúa Villaseñor examines how the stigmatization of an urban district can be understood as a two-directional process; one in which certain areas become more upscale and chic through the demonization of others. Relying on approaches from Latin American Cultural Studies, he works out to what extend the media's gaze at no-go areas reinforces the urban segregation and urban imageries of fear on the one hand, and the chic and safe nature of other districts on the other hand.

Selma Siew Li Bidlingmaier's chapter "Spaces of Alterity and Temporal Permanence" explores the means by which the representation and popular perception of Chinatowns in the United States as exotic, dangerous, mysterious, morally depraved, and outlandish have been produced, maintained, and neutralized over the past two centuries. Drawing from Michel Foucault's principle of heterochrony, Bidlingmaier argues that Chinatowns are counter-sites within cities that encapsulate an "absolute break with traditional time." These "frozen" spaces become museums of an "immutable Orient," preserved and untouched by the progression of time and civilization development. With Chinatowns hauling in billions of dollars from tourism each year, these myths become part of a complex system of economic, social, and political relations which the chapter aims to deconstruct.

Works Cited

Chrish, K. 2010. *Hyper Ghetto*. [Online]. Available at: www.hyperghetto.com [accessed: June 15, 2010]

Huggan, G. 2001. *The Postcolonial Exotic: Marketing the Margins*. London: Routledge.

Lash, S. and Urry, J. 1994. *Economies of Signs and Space*. London, Thousand Oaks, and New Delhi: Sage.

Maffesoli, M. 1996. *The Time of the Tribes: The Decline of Individualism in Mass Society*. London: Sage.

Sassen, S. 1998. *Globalizatins and Its Discontents*. New York: New Press.

Smith, N. 1987. Gentrification and the rent-gap. *Annals of the Association of American Geographers*, 77(3), 462–65.

Smith, N. 1996. *The New Urban Frontier. Gentrificatrion and the Revanchist City*. London and New York: Routledge

Villa, R.H. 2000. *Barrio-Logos: Space and Place in Urban Chicano Literature and Culture*. Austin: University of Texas Press.

Wacquant, L. 2007. *Urban Outcasts: A Comparative Sociology of Advanced Marginality*. Cambridge: Polity Press.

Yúdice, G. 2003. *The Expediency of Culture*. Durham, NC: Duke University Press.

Zapf, H. 2002. *Literatur als kulturelle Ökologie: Zur kulturellen Funktion imaginativer Texte an Beispielen des amerikanischen Romans.* Tübingen: Niemeyer.

Zukin, S. 1987. Gentrification: culture and capital in the urban core. *American Review of Sociology*, 13, 139–47.

Zukin, S. 1989. *Loft Living: Culture and Capital in Urban Change.* New Brunswick: Rutgers University Press.

Zukin, S. 1995. *The Cultures of Cities.* Oxford and Malden, MA: Blackwell.

Chapter 13

(Re-)Constructing the Ethnic Neighborhood: Gentrification in the United States and the Longing for a Unique Identity[1]

Eric C. Erbacher

In the past decades, gentrification has become one of the most vigorously discussed phenomena of the post-Fordist city. Conventionally defined as "the restoration and upgrading of deteriorated urban property by middle class and affluent people, often resulting in displacement of lower-income people" (Gentrification 2006), gentrification processes have visibly affected and transformed many inner-city neighborhoods across the United States. While the term's connotations of class takeover of the inner city initially led to various attempts by supporters of the process to dispose of the "dirty word" and replace it with more palatable labels such as "urban renaissance," "neighborhood reinvestment," or "inner-city revitalization," both the term "gentrification" as well as the various processes connected to it have become the norm in contemporary urban development and in its discussion (Lees, Slater, and Wyly 2008: 154–8). Particularly in the context of the general restructuring from an industrial to a postindustrial economy, the ascendancy of the knowledge and information sectors, and an ensuing intense urban competition for a "creative class" (Florida 2005) as the supposed guarantor of a city's economic survival, the presence and production of gentrified spaces has become an important asset and a "global urban strategy" (Smith 2002) for various private and public actors in the urban realm. Typically, neighborhoods experiencing gentrification exhibit a wide variety of social and ethnic, but also physical and architectural diversity. Although this heterogeneity might seem to limit the possibilities of creating easily consumable themed urban spaces, the neighborhoods' diverse history and populations are able to exactly provide what large segments of the "creative class" seem to yearn for: presumably authentic and unique experiences within a neo-bohemian lifestyle (Lloyd 2006). Particularly a neighborhood's ethnic past and the presence of ethnic groups have often come to play a significant role in attempts to promote and market the neighborhood's lifestyle options.

1   I would like to thank the German Academic Exchange Service (DAAD) for a grant that partly supported the research for this article.

This chapter will show how representations and constructions of ethnicity in print media contribute to gentrification processes by presenting particular aspects of a neighborhood's ethnicity as conducive to the fulfillment of the yearning for diversity and authentic experiences, even if these are packaged as tamed commodities. In the first part, the focus will be on the origins and the development of the potential gentrifiers' value structures and the function that ethnicity fulfills within these structures.[2] The second part will deal with mass media representations and their capacity to construct collective spatial images and associations that support the marketing and selling of certain neighborhoods. The third part will take a closer look at how certain ethnic elements are represented and constructed as desirable by local print news media in their coverage of the neighborhoods of Wicker Park, Chicago, and Williamsburg in Brooklyn, New York, that have recently undergone gentrification.

The media analysis employed here is based on the qualitative discursive analysis of local mass-circulation newspapers from both cities from 1980 to 2005, a period of rapid and intense gentrification in both neighborhoods. The individual newspapers were chosen as to represent a wide range of depictions of the process and the changes in the neighborhoods and thus include a daily quality newspaper, an alternative weekly magazine as well as a monthly city magazine. Despite the variety of media included, the focus here is on media with a circulation large enough to create the socially powerful collective spatial images relevant to this chapter. Small or highly local publications that might deal with neighborhood change and the gradual displacement of ethnic populations much more critically are thus not considered here, which also accounts for the scarcity of more outspoken countering voices.

Ethnic Diversity: An Important Part of the Gentrifiers' Value Structure

The search for exciting experiences is part of wider changes in consumer culture that emerged parallel and in close connection to the post-Fordist economy (Featherstone 1991). As the fairly clear-cut social structure of the old Fordist system based on occupational standing eroded due to the growth and diversification in white-collar professional employment, giving rise to what has been termed a "new middle class" (Ley 1996), consumption emerged as an important, maybe the most important, marker of social position. Accompanied

---

2   The term gentrifiers will here be used to refer to the incoming population groups, many of whom belong to the "new middle class" or the "creative class" and are in their majority, although by far not exclusively, white Anglos. In contrast, the term ethnic groups is employed as a catch-all term to indicate non-Anglo population groups of various ethnic origins, particularly those who have been living in the respective neighborhood for longer periods of time.

by a production crisis of mass-produced consumer goods targeted at mass markets, the new neoliberal system of flexible production and niche marketing provided a plethora of consumption choices as a means of social differentiation as well as class constitution for this diverse new social group (Harvey 1992). While consumption as a strategy of stratification had already been described by sociologist Thorstein Veblen in 1899, the new middle class's focus on lifestyle and experiences is dependent on new definitions of what constitutes social and cultural capital, and thus economic values. Instead of trying to keep up with the "Joneses" and their proverbial suburban lifestyle, many middle-class consumers growing up in the countercultural atmosphere of the 1960s and 1970s wanted to be different from them (Rutherford 1990: 11). Influenced by celebrations of diversity, creativity, tolerance, and individuality and a resulting general rejection of homogeneity, uniformity, and mass production and consumption, many members of the educated middle class, whose material needs had largely been fulfilled, turned to culture and lifestyle issues to construct their social identity. According to sociologist Mike Featherstone, it was "a movement away from agreed universal criteria of judgment of cultural taste towards a more relativistic and pluralistic situation in which the excluded, the strange, the other, the vulgar, which were previously excluded [could] now be allowed in" (1991: 106).

Thus one way to perform one's individuality is through the consumption of niche and specialized consumer goods increasingly marketed as being "different," "ethnic," or imbued with an image of rebellion, as described by Thomas Frank in his pointed analysis of advertisement's appropriation of bohemian lifestyles:

> In contemporary American public culture the legacy of the consumer revolution of the 1960s is unmistakable. Today there are few things more beloved of our mass media than the figure of the cultural rebel, the defiant individualist resisting the mandates of the machine civilization. ... [R]ebellion is both the high- and the mass-cultural motif of the age. (1997: 228–9)

A product's image is thus at least as important for the purchase as the use value itself. Choice of residential location, however, seemed to offer an even more effective affirmation of avantgarde status for those seeking to distinguish themselves from the traditional middle classes as well as from their less fashionable peers. As suburbia represented almost everything that this new group scorned and rejected—social, ethnic, and aesthetic homogeneity as well as low density, mass production, and social control—many turned towards the old, sometimes decaying, ethnically and socially diverse and dense inner-city neighborhoods, thereby initiating gentrification processes.

Research on gentrification has long been divided into two different approaches: Production side interpretations, such as those by Marxist geographer Neil Smith, have regarded gentrification as the result of uneven development and capital's search for undervalued urban real estate and have thus contributed to our understanding of the structural forces behind the process (1996). Demand side explanations of gentrification, however, shift the focus on the gentrifiers, their motivations, and the cultural and social origins of these motivations. They therefore not only help explain the very powerful aesthetic and sensual experience of gentrification but also restore meaningful agency to the gentrifiers and their actions. In this vein, gentrification can be regarded as an "attempt to recapture the value of place" (Zukin 1991: 192) by individuals of a social group that wants to exhibit its liberal tolerance and cultural sensibility and refinement against suburban conformity and kitsch. Emphasizing lifestyle enables the generally well-educated yet frequently not wealthy gentrifiers to construct "a social space or *habitus* on basis of culture rather than economic capital" (Zukin 1991: 192 original emphasis) Echoing Pierre Bourdieu's notion of habitus as a means of establishing and performing social distinction (1986), the gentrifiers' sociospatial habitus is very much dependent on a particular lifestyle, which, in turn, is contingent on a particular place—the place of the inner city with its cultural connotations of difference, diversity, ethnicity, authentic experiences, and experimentation. In order to make these highly regarded values operational to identity construction, they must be acquired and appropriated. This appropriation must take place in a manner that signifies the inmovers' social and cultural difference and superiority to the working- and lower-class populations (stereo-) typically living in and associated with the inner city. The highly reflective and selective consumption of certain places and experiences that denote both sophistication as well as seemingly authentic simplicity makes it possible to project an image of a refined and discerning yet curious and unpretentious personality (Zukin 1991: 202).

Next to places and situations promising the experience of history, artistic cultures, or bohemia (Zukin 1989, Zukin 2010, Matthews 2008), ethnicity and ethnic diversity are highly sought after, marketed, and consumed characteristics of gentrified neighborhoods. Many neighborhoods undergoing gentrification have been, and still are, home to various ethnic groups and usually exhibit numerous traces of ethnic life and culture. Yet, the influx of typically white Anglo members of the "creative class" poses a threat to the often low-income ethnic and minority groups living, working, and socializing in the neighborhood. Even though recent studies have called into question assumptions of a large extent of direct displacement (Freeman 2006, Vigdor 2002), the increase in property value followed by rising rents and property taxes caused by gentrification usually makes it difficult or impossible for many long-term, ethnic inhabitants on a fixed income to stay in a gentrifying neighborhood and benefit from the area's usual improvements such as declining crime, better municipal services, and more shopping opportunities. At the same time as many neighborhoods undergoing gentrification witness the opening of the first supermarket in many years (Williams 2008), the gentrifiers' selective consumption choices often spell the eventual end of ethnic businesses or gathering spots considered too ethnic or lower-class by the gentrifiers. In this way, the ethnic community is frequently deprived of long-standing socially and culturally important network ties, leading to what urban geographer Mark Davidson has called "neighborhood resource displacement:"

Neighbourhood resource displacement involves the changing orientation of neighbourhood services and the increasing "out-of-placeness" of existing residents. It recognises that, as a neighbourhood transitions, not only does the neighbourhood social balance change, but also local shops and services change and meeting-places disappear. The places by which people once defined their neighbourhood become spaces with which they no longer associate. (2008: 2392)

Although some ethnic communities form organizations in an attempt to counter and resist the transformations in their neighborhood and are frequently successful in at least softening direct displacement pressure,[3] they are generally unable to stop or even reverse their community's increasing economic, political, social, and cultural marginalization. In a vigorously argued study of the West Town area of Chicago, which includes the neighborhood of Wicker Park to be discussed below, John J. Betancur (2002) has shown how ethnic organizations fighting for low-income housing have increasingly been sidelined in community affairs by the politically, economically, and socially better connected gentrifiers and their efforts at neighborhood transformation.

While harmful to the affected ethnic community, the resulting loss of ethnic and social diversity eventually also poses a danger to the neighborhood's attractiveness to residents, visitors, and tourists. For present and prospective gentrifiers, the increasing homogeneity of their neighborhood threatens to disrupt their spatial claims to an exciting inner-city lifestyle full of "authentic" ethnic experiences, thereby putting both their cultural and economic capital at risk. For visitors and tourists drawn to the ethnic heritage and supposedly genuine streetscapes with old shops and "real" ethnic restaurants, the disappearance of these seemingly rooted places and their replacement with often generic sites of upscale consumption, might lead to a loss of interest and a redirection of their money flow to other areas. Such a drop in marketable uniqueness also represents a threat to boosterist municipal agencies, real estate developers, and local media hoping to project an image of a lively and exciting city with distinctive and fashionable neighborhoods in their interurban competition for members of the affluent and supposedly economically vital "creative class." Thus, almost all affected parties, from residents and visitors to public officials and the local business community, seem to have a stake in maintaining and promoting the gentrifying neighborhood's image as a place where a rich ethnic past is combined with a vibrant and authentic ethnic present. An essential part of this image are the many ethnic businesses which have survived the loss of their original customer base by adapting to the gentrifiers' and visitors' tastes and have come to rely on the steady stream of income from these customer groups.

The neighborhood's ethnic past is therefore sustaining its ethnic image in the present, which, in turn, helps to support at least some remaining ethnic places.

---

3   For examples of community opposition and resistance strategies see Lees, Slater, and Wyly 2008, 246–276.

While the perceived authenticity of the ethnic experience is certainly a highly valuable commodity, its more unruly and less desirable aspects need to be controlled in order to be palatable to an audience demanding Anglo middle-class behavioral norms and security. In this way, some common ethnic social and cultural practices like sitting on the stoop or large and exuberant gatherings of friends and relatives are considered almost as much examples of deviant behavior as social problems more traditionally associated with marginalized low-income ethnic communities such as gang crime, drug dealing, or alcoholism. The more a neighborhood becomes gentrified, the more not only are the criminal activities policed, but also, as Betancur has shown, are those practices controlled that are at variance with the gentrifiers' notion of correct behavior (2002).

Mass Media Representations and their Capacity to Construct Marketable Neighborhood Images

In the gentrification of ethnic neighborhoods, the media play an important role in shaping the neighborhood's image by both highlighting and promoting its desirable features as well as by moderating, framing, and downplaying the more problematic ones—all of this from a middle-class and urban boosterist perspective. Largely independent of the media's political outlook, this perspective is mainly necessitated by economic considerations. Mass-market local news media are to a great extent economically dependent on two local communities: A sizable, affluent, and consumption-oriented, that is middle-class, readership as well as a strong and prosperous local and regional business community. The former provides income not only through individual sales or subscriptions, but, much more importantly, by producing the circulation figures needed to generate advertising revenue from the companies promoting their products to middle-class readers (Bagdikian 2004). In a society where homeownership is not only highly valued culturally but also the basis of large amounts of its financial equity, real estate agents and developers account for an important part of the advertising revenue. Apart from immediate economic concerns, local media are also reliant on close connections to municipal agencies for access to information on local developments and affairs. Mass-market local news media are therefore contingent on a large middle-class audience as readers and customers for the advertised products, on thriving businesses for advertising income, especially from the real estate sector, as well as on public officials and municipal bodies. Thus being part of what geographers Elvin Wyly and Daniel Hammel have called "urban growth regimes" (2008: 2647), local media have converging interests and frequently close cooperations with boosterist municipal agencies and large real estate developers in attempts to attract affluent newcomers to their city. While these connections are undoubtedly influential, this is not to imply that contemporary gentrification processes in ethnic neighborhoods are single-handedly engineered and controlled by local news media. Media representations, however, help middle-class consumers become familiar with and appreciate certain ethnic characteristics of a neighborhood by constructing its image in a way that alludes to their countercultural sensibilities while also accommodating to their anxieties.

This power of the media over the imagination of the consuming public can be exercised in several ways: According to Maxwell McCombs and Donald Shaw, by choosing to present a topic in a certain way, the media, whether wittingly or unwittingly, are able to set the agenda for an issue:

> Whatever the attributes of an issue—or other topic—presented on the news agenda, the consequences for audience behavior are considerable. How a communicator frames an issue sets an agenda of attributes and can influence how we think about it. Agenda setting is a process that can affect both what to think about and how to think about it. (1993: 62–3)

Moreover, in the case of ethnic neighborhoods, the culturally induced attributes that journalists employ when describing certain elements of a neighborhood's ethnic past and present as fashionable or desirable are a reflection of their own cultural and social sensibilities as members of the creative class and provide an "aesthetic critique that facilitates upscale consumption" (Zukin 1991: 202) for culturally similarly disposed audiences. Journalists are thus part of what Zukin (1991: 258) has called gentrification's "critical infrastructure:" gentrifiers in creative professions such as chefs, gallery owners, and writers catering to the needs of their own social group and thereby shaping its tastes and predilections. In the present context, this involves the "aestheticization of diversity" (Zukin 1995: 2). This exemplification of an ethnic neighborhood's attractions is particularly important and effective for audiences without personal knowledge of the respective area.

The ways that ethnic places are represented play a crucial role for the image of the neighborhood as a whole. Urban sociologist Christopher Mele points out that

> representations "stand for" a place and offer an alternative to immediacy and experience as a means to know about a neighborhood and its residents. The range of particular images, symbols, and rhetoric that visitors, observers, and outsiders use to refer to or describe a neighborhood as pleasant or dangerous are constructed neither necessarily by the gazers themselves nor by those with firsthand knowledge through use of such spaces. Instead, people come to know of and about such places through the circulation of predominant representations and popular notions. … [R]epresentations are hegemonic processes that tend to pervade the everyday consciousness of non-residents and residents alike. (2000: 13–15)

Moreover, through constructing and maintaining representations of ethnic places as both authentic and safe, the media are able to authenticate certain ethnic activities, lifestyles, and places while simultaneously providing instructions on how to experience and enjoy them in a secure environment. Similar to the way that common ethnic social practices like outdoor socializing are considered deviant and are becoming increasingly policed, characteristics of ethnic neighborhoods that might be potentially disturbing to middle-class consumers such as destitute living and working conditions or overly exotic food are progressively being displaced from the representation. Ethnicity thus becomes a tamed commodity.

Taken together, media coverage of gentrifying ethnic neighborhoods constructs symbols, images, and a rhetoric that are part of a cultural and social discourse closely connecting inner-city spaces with valorized ethnic elements resulting in an increased demand of middle-class consumers for these semantically loaded places. This adds up to a possibly unconscious, but certainly consented to, selling of the ethnic inner city which will be exemplified with the following examples of the local news media coverage of two neighborhoods undergoing gentrification, Wicker Park, Chicago, and Williamsburg in Brooklyn, New York City.

Constructing Wicker Park and Williamsburg as Ethnic Consumption Places

Chicago's Wicker Park is an inner-city neighborhood that was developed in the mid- to late 1800s by middle-class German and Scandinavian traders and craftsmen such as brewers and cabinetmakers, who built many of the still existing, elaborate Victorian mansions for themselves and workingmen's cottages and row houses for their workers in the nearby factories. Around the turn of the century, these original owners were slowly succeeded by Polish immigrants who, in turn, gave way to a Jewish, African American, and Hispanic—mostly Puerto Rican—population, each leaving their marks on the neighborhood's physical and cultural fabric. In the early 1980s, white middle-class singles and couples began to move into the physically and financially neglected, yet leafy, lively, and diverse area close to the central business district and started to engage in low-scale and private rehabilitation work on the old mansions and row houses (Lloyd 2006: 27–46). As the neighborhood was still being mainly inhabited by a variety of ethnic minorities, Chicago's print media focused on Wicker Park's ethnic diversity in their coverage. They continuously stressed especially the positive and attractive qualities of the ethnic amenities, as this boosterist quotation from the May 1980 issue of *Chicago Magazine* shows:

> Wicker Park: Boosters of this Near Northwest Side community say it's perfect for suburbanites or high-rise dwellers who are tired of architectural and ethnic homogeneity. Everything is not as spic and span ... but there are a number of advantages: eight minutes to the Loop by rapid transit, a diversity of people, each of whom has dotted the community with distinctive restaurants and stores; and fine old mansions that date from the time when Wicker Park was the ethnic millionaires' Gold Coast. (Grossman 1980: 156)

Possibly even more effective in creating a favorable image of Wicker Park's ethnic diversity are these passages from a 1983 *Chicago Tribune* interview with a young Anglo couple living in a rehabilitated house:

> Bill: It [Wicker Park] has roots. ... There's an ethnic diversity, almost a celebration of it. Kathryn: I've met more interesting people in the four years we've lived here than I've met in a number of years. ... Bill: I want ethnic diversity. ... [W]e found that moving here had a great liberating effect. ... This is not to spit on the suburbs at all, but I find nothing at all that I miss. (Schilling 1983: A3)

Both passages, which are representative of many articles on Wicker Park during the early 1980s, describe ethnic diversity as a desirable general characteristic of the neighborhood, touting it as a central facet of an exciting inner-city lifestyle that is consumable through residence in the "fine old mansions" of the "ethnic millionaires," by visiting the "distinctive restaurants and stores," and by the possibility of meeting "interesting people." Moreover, Wicker Park is presented as "perfect for suburbanites or high-rise dwellers who are tired of architectural and ethnic homogeneity," hinting at means of differentiation and distinction for newly middle-class readers. Wicker Park's ethnic populations thus became an essential facet in the medial positioning of Wicker Park as the antithesis of bland suburbia.

As spreading gentrification in the late 1980s and early 1990s resulted in a continuous influx of middle-class Anglo-Americans and a reduction of Wicker Park's ethnic diversity, media coverage underwent a slight shift. On the one hand, the media welcomed the changes to the neighborhood's socioeconomic structure as signs of its long-awaited revitalization; on the other hand, they routinely described the visibly threatened social and ethnic characteristics responsible for the area's attractiveness as not being lost yet and still being present and waiting to be experienced, as in this quotation from an October 1992 *Chicago Tribune* article:

> There's still an edgy quality here, a tension in the air, perhaps a product of the forced interaction of longtime white ethnic, Hispanic and black residents and the more-recent arrivals—artists initially attracted by cheap rents and professionals who fled the crowded lakefront. ... "Wicker Park has its own vibe," contends Tom Handley, owner of [a local coffeehouse] Urbus Orbis. ... "A big part of the original community never left. There's still a large Ukrainian, Polish and Czech neighborhood here. And most of the people who have come to Wicker Park in the latest wave, including me, really like it the way it is." (Lauerman 1992: 12)

Yet, despite these evocations of its ethnic and social diversity and the pleas for their preservation, as disingenuous as they may be, the neighborhood is increasingly becoming more ethnically homogeneous. According to United States census data, Hispanics and Latinos, for instance, accounted for just over 50 per cent of Wicker

Park's population in 1990, whereas their proportion had dropped to 28 per cent only a decade later (Lloyd 2006: 115).

While the history of the various ethnic groups in the neighborhood had long been a feature of many articles, the decrease in non-Anglo ethnic diversity in the early 1990s seemed to prompt newspapers to stress their remaining traces. The focus was generally put on the two most sizable ethnic populations in Wicker Park: the Polish and the Latino communities. In the cases of both communities, their traditional foodways as preserved in a number of restaurants emerged as consumable links between Wicker Park's history as an ethnic neighborhood and its gentrified present. As other means of experiencing Wicker Park's ethnic communities became scarcer, visiting a restaurant represented as authentically ethnic was framed as a possibility to appropriate the neighborhood's history in a presumably safe atmosphere. This connection becomes clear in the following quotation from a September 1994 *Chicago Reader* article:

> [Wicker Park's] original settlers were German and Scandinavian, but they left no remnants as they fled the Poles, who made this the heart of Chicago's Polonia for more than half a century—though there were Ukrainian, Italian, Serbian, and Eastern European Jewish enclaves within it. Later the area became predominantly Latino—mainly Puerto Rican with some Mexican—as many Euros fled. And recently artists and yuppies began to fill in a lot of the spaces. Today there's still a great little Polish American diner at the confluence of Wicker Park and Bucktown, the legendary Busy Bee ..., where you can get everything from pierogi to a hot meatloaf sandwich. ... The most exotic of the neighborhood's Latino restaurant is Rinconcito Sudamericano ..., offering the many magical potato dishes and seafoods of Peru. Lydia's Cafe ... is basically Puerto Rican, an amiable spot located in the back of a small grocery that features good black beans and other standards. Well-prepared traditional Mexican dishes are served at Tecalitlan ... in a pleasant dining room adjacent to the take-out section where you enter. The Mexican seafood at Costa Azul ... ranges from the enormous shellfish cocktails to the lusty, garlic-glazed snapper. (Rose 1994: 7)

The prevalent detailed descriptions of ethnic restaurants, of their interiors, their owners and staff, and also of the food served arguably helps to both authenticate the ethnic dining experience as well as to prepare prospective middle-class (and) Anglo customers for their exposure to potentially overly exotic and disturbing environments and foods. The consumption of both the ethnic places as well as their food is thus framed as a safe yet authentic experience, given character and made even more intriguing through the use of adjectives such as "legendary," "exotic," "magical," and "lusty."

As gentrification proceeded to transform Wicker Park into a solidly middle- and upper-class neighborhood with a majority Anglo population, a growing number of more recent ethnically themed restaurants with often dubious claims to local community roots, such as the Mexican-themed chain restaurant Adobo Grill, joined the few remaining traditional ethnic restaurants (Knab 2005). Together, they are increasingly burdened with providing and maintaining Wicker Park's image of a historically ethnic neighborhood offering consumable ethnic experiences. A case in point is the above-mentioned Polish diner Busy Bee, which had to represent Wicker Park's Polish past almost single-handedly in its heavy coverage in Chicago's newspapers and magazines from the mid-1980s to its eventual closing in 1998, including cover articles several pages long. On the much-lamented occasion of its closing, the *Chicago Tribune* celebrated it as a:

> haven ... since the late 1960s, when the Busy Bee depended on the solidly Polish ethnic base of Wicker Park for its business, to its vaunted status today as a city institution and popular stumping ground for politicians. ... On June 28, the 69-year-old Polish immigrant [and owner Sophie Madej] will serve up her billionth and last pierogi before kicking out the regulars—from the local cops to the self-proclaimed "Over-the-Hill Gang" of seniors to the new crew of pierced and tattooed twentysomethings. (Ryan 1998: 5)

At the turn of the twenty-first century, many of Wicker Park's old ethnic restaurants and meeting places have thus disappeared along with the majority of its ethnic population sustaining them. Next to themed restaurants, it is largely left to promotional street signs identifying the neighborhood's formerly predominant ethnic communities, the Polish and Hispanic communities, to uphold the memory of their presence in the neighborhood. Wicker Park's successful depiction as an ethnically diverse and rich area and its marketing to a newly middle-class clientele with a thirst for an experiential lifestyle in an ethnic neighborhood thus has eventually resulted in the loss of many of its actual ethnic elements and their replacement with a largely symbolic historical ethnicity that can safely be appropriated through supposedly "authentic" restaurants or formerly ethnic-inhabited real estate.

While large parts of Wicker Park's ethnic communities have been displaced or marginalized, Brooklyn Williamsburg's ethnic groups have maintained a much more visible presence, even though they too have been used for the neighborhood's promotion to middle-class home buyers. The old industrial and working-class neighborhood of Williamsburg in Brooklyn, located directly across the East River from Manhattan, became a desirable residential location for many immigrants during the course of the twentieth century, primarily due to its abundance of employment opportunities. Initially settled by upper middle-class Anglos as a suburb in the 1830s, by the turn of the twentieth century the neighborhood's population were mostly Polish Americans and Italian Americans. In the 1940s and 1950s, Orthodox Hasidic Jews of the Satmar community escaping persecution in Europe moved to Williamsburg, eventually forming a sizable and politically influential enclave. This ethnic diversity was added to, in the 1960s and 1970s, with the influx of Hispanics, mostly Puerto Ricans. Yet, just as Williamsburg became a destination for new immigrants, most of the large industrial employers closed down or relocated, leaving large parts of the population to suffer the effects of unemployment, communal neglect, discrimination, and often ensuing social problems.

However, by the late 1980s, Williamsburg was "discovered" first by artists and, during the 1990s, by more conventionally employed members of the "creative class" (Curran 2004: 1243–50). Similar to Wicker Park, a celebration of its ethnic diversity accompanied the early stages of the gentrification process in Williamsburg, as in this passage from a 1992 *New York Times* article:

> Few neighborhoods are better illustrations of [mayor] Mr. Dinkins's "gorgeous mosaic" than Williamsburg. The [Orthodox Jewish] Hasidim have recreated something of a shtetl on the blocks around Lee Avenue. The signs are mostly in Yiddish, the smells are of babkas and stuffed cabbage. But woven in and around this enclave are restaurants fragrant with arroz con pollo and lusty with the sounds of salsa. ... [T]here is something life-affirming and lyrical in a tenant roster, like the one in a local public housing project that contains this sequence of names: "Garcia, Goldstein, Gonzalez, Greene, Gross, Guttman." (Berger 1992: B3)

Once again, ethnic diversity appears here as both a positive neighborhood asset and an intrinsic part of its identity. Similarly to Wicker Park, detailed description of places and (stereo-)typical ethnic foods, which can be experienced and appropriated through consumption in ethnic restaurants framed as authentic, transport much of Williamsburg's ethnicity. In the above quotation, this is achieved through the use of foreign-language terms such as "shtetl," "babkas," or "arroz con pollo," where the lack of translations or explanation presupposes a comparatively knowledgeable reader.

As its gentrification advanced in the 1990s, some of Williamsburg's ethnic groups such as the Puerto Rican population and the Satmar Hasidim could uphold their heavy presence in the neighborhood, while others like the Italian and Polish communities largely moved to other parts of New York City. Yet, as in Wicker Park, medial representations of Williamsburg drew heavily on the ethnic history and present, if available, which they then employed to imbue the Williamsburg of the late twentieth and early twenty-first centuries with the aura of a long tradition of ethnic groups living and leaving their traces in the area. Ethnic elements, whether visibly present in the neighborhood or only discursively attached to it, thus became ready to be incorporated into a middle-class newcomers' lifestyle in search of roots and difference. The following quotation from the *New York Times* of 2003 exemplifies this:

> It is the Williamsburg [Bridge] that opened Brooklyn to the proletarian Jewish and Italian immigrants who had been crowded into the ghettoes of the Lower East Side, shaping the Brooklyn whose earthy charm endures today. ... In

serving as this gateway, the bridge created some of New York's most flavorful neighborhoods. There are the streets filled with Yiddish-speaking Hasidim branching off clamorous Lee Avenue, the awning-shaded row houses and indolent cannoli cafes in the Italian enclave to the north, the domino players and Technicolor murals of the Puerto Rican Southside, the garage art galleries of the bohemian quarter along Bedford Avenue. ... While residents of other Italian neighborhoods have fled to suburbia, those of Williamsburg have stayed close to the bridge. Joseph Sciorra of the John D. Calandra Italian American Institute at Queens College, has lived on Devoe Street for 20 years and likes a place where he can have a fine meal at S. Cono's Pizzeria or pastries at Fortunato Brothers, started in 1976 by four brothers from a village near Nola. "I like the fact that it's still a neighborhood," Mr. Sciorra said. "If I don't have money on me at the local grocery store, the guy gives me credit till I come back next time. If I've forgotten something upstairs, I can leave my kids on the stoop and not worry." (Berger 2003: 25)

While the article also mentions, as if to establish common ground with the readers, the ethnic groups that are known to call Williamsburg home, Orthodox Satmar Jews and Puerto Ricans, the focus in the above quotation is on the neighborhood's Italian community, which is not as widely associated with Williamsburg. Its representation as a traditional part of Williamsburg's "gorgeous mosaic" thus seems to be even more important as it adds another ethnic consumption option for potential residents. The article frames Williamsburg's Italian community as being more authentic than other Italian communities and as having preserved seemingly lost community values such as trust and mutual support—thus feeding a middle-class nostalgia for a supposedly better past that can be alleviated by buying into the neighborhood. Also, the detailed description of the food places and the personalization of the narrative validate and authenticate Williamsburg as the location of supposedly "real" ethnic experiences.

Adding to the image of Williamsburg as the locale of a tightly knit traditional Italian community are the repeated reports on its yearly Giglio festival, as in this exemplary quotation from the *New York Times*:

An ornate 85-foot tower, festooned with papier-mache lilies, angels and cherubim, bobbed and swirled through the Williamsburg section of Brooklyn yesterday afternoon as about 30,000 people crowded onto the normally quiet blocks around North Eighth and Havemeyer Streets for one of New York's oldest and liveliest festivals. The occasion was the annual dancing of the giglio—a mid-July rite since 1903 in this working-class Italian-American neighborhood that combines expressions of religious devotion and community solidarity with the display of one of the city's quirkiest works of folk art and a dose of marching machismo. (Yarrow 1990: B3)

In the context of gentrification, ethnic festivals like the Giglio arguably fulfill a double function: On the one hand, for the ethnic communities, they act as meaningful opportunities to express their culture and to lay claim to certain urban spaces, even if only symbolically. On the other hand, ethnic festivals are able to combine and offer many of the desired characteristics that help to make ethnic neighborhoods so attractive to visitors, tourists, and home buyers: a combination of people, history, tradition, food, and spectacle, all presented in a seemingly authentic yet stereotypical ethnic package and grounded in a particular place loaded with ethnic experiences as easily consumable as a trip to the neighborhood where they are medially located. For those more deeply concerned with their social and cultural status as members of an open-minded, diversity-loving, difference-seeking class, the gentrified ethnic neighborhood as presented, authenticated, and secured through media discourse offers a possibility to purchase a fashionable lifestyle through the buying of gentrified inner-city real estate.

Represented by media coverage as agents of an experiential lifestyle or exciting Otherness, a neighborhood's ethnic history, its ethnic communities, and its ethnic traditions can thus unwittingly contribute to the attractiveness of an area, accelerate and define its gentrification, and aid in its marketing to potential gentrifiers. While ethnic groups might also use media exposure and gentrification processes to their advantage, the commodification of their place of residence as an ethnic neighborhood is ultimately not dependent on their physical presence. As real estate becomes too valuable for their actual presence, symbolic references are often enough to sell the ethnic neighborhood.

Works Cited

Bagdikian, B. 2004. *The New Media Monopoly*. Boston: Beacon Press.

Berger, J. 1992. Metro matters: fear builds a bridge across gulf of cultures. *New York Times*, August 31, B3.

Berger, J. 2003. The other bridge, but all Brooklyn: over 100 years, the Williamsburg has diversified a borough. *New York Times*, June 22, 25.

Betancur, J. 2002. The politics of gentrification: the case of West Town in Chicago. *Urban Affairs Review* 37(6) 780–814.

Bourdieu, P. 1986. *Distinction: A Social Critique of the Judgment of Taste*. London: Routledge.

Curran, W. 2004. Gentrification and the nature of work: exploring the links in Williamsburg, Brooklyn. *Environment and Planning*, 36(7), 1243–58.

Davidson, M. 2008. Spoiled mixture: where does state-led "positive" gentrification end? *Urban Studies*, 45(12), 2385–2405.

Featherstone, M. 1991. *Consumer Culture and Postmodernism*. London: Sage.

Florida, R. 2005. *Cities and the Creative Class*. New York and London: Routledge.

Frank, T. 1997. *The Conquest of Cool: Business Culture, Counterculture, and the Rise of Hip Consumerism.* Chicago: University of Chicago Press.

Freeman, L. 2006. *There Goes the 'Hood: Views of Gentrification from the Ground Up.* Philadelphia: Temple University Press.

Gentrification. 2006, in *The American Heritage Dictionary of the English Language.* 4th ed. Boston: Houghton Mifflin.

Grossman, R. 1980. Beyond the Lakefront. *Chicago Magazine*, May, 156.

Harvey, D. 1992. *The Condition of Postmodernity.* Oxford: Blackwell.

Knab, A. 2005. Buenos dias! Clink drinks to Dia de los Muertos with tequila, mescal and more. *Chicago Tribune*, October 25, 21.

Lauerman, C. 1992. Growing pains: new money and new residents are changing the character of Bucktown and Wicker Park: not everyone thinks the changes are for the better. *Chicago Tribune*, October 18, 12.

Lees, L., Slater, T. and Wyly, E. 2008. *Gentrification.* New York and London: Routledge.

Ley, D. 1996. *The New Middle Class and the Remaking of the Central City.* Oxford: Oxford University Press.

Lloyd, R. 2006. *Neo-Bohemia. Art and Commerce in the Postindustrial City.* New York and London: Routledge.

McCombs, M. and Shaw, D. 1993. The evolution of agenda-setting research: twenty-five years in the marketplace of ideas. *Journal of Communication*, 43(2), 58–67.

Matthews, V. 2008. Artcetera: narrativising gentrification in Yorkville, Toronto. *Urban Studies*, 45(13), 2849–76.

Mele, Ch. 2000. *Selling the Lower East Side: Culture, Real Estate, and Resistance in New York City.* Minneapolis and London: University of Minnesota Press.

Rose, D. 1994. Restaurant tour: dining around the Coyote. *Chicago Reader*, September 9, 7.

Rutherford, J. 1990. A place called home: identity and the cultural politics of difference, in *Identity: Community, Culture, Difference*, edited by J. Rutherford. London: Lawrence and Wishart, 9–27.

Ryan, N. 1998. Busy bee to buzz no more: a hive of activity for several decades, the Wicker Park Polish diner that served kolaczkis and pierogis to cops and politicians is closing. *Chicago Tribune*, June 6, 5.

Schilling, T. 1983. Dialogue: ex-suburbanites find city life liberating. *Chicago Tribune*, January 7, A3.

Smith, N. 1996. *The New Urban Frontier: Gentrification and the Revanchist City.* London and New York: Routledge.

Smith, N. 2002. New globalism, new urbanism: gentrification as global urban strategy. *Antipode*, 34(3), 434–57.

Vigdor, J. 2002. Does gentrification harm the poor?, in *Brookings-Wharton Papers on Urban Affairs 2002*, edited by W. Gale and J. Rothenberg Pack. Washington, DC: Brookings Institution Press, 134–73.

Williams, T. 2008. Powerful Harlem church is also a powerful Harlem developer. *New York Times*, August 18, B1.

Wyly, E. and Hammel, D. 2008. Commentary: urban policy frontiers. *Urban Studies*, 45(12), 2643–8.

Yarrow, A. 1990. Joy and challenge mix in dancing of the Giglio. *New York Times*, July 9, B3.

Zukin, S. 1989. *Loft Living: Culture and Capital in Urban Change.* New Brunswick, NJ: Rutgers University Press.

Zukin, S. 1991. *Landscapes of Power: From Detroit to Disneyworld.* Berkeley et al.: University of California Press.

Zukin, S. 1995. *The Cultures of Cities.* Cambridge, MA: Blackwell.

Zukin, S. 2010. *Naked City: The Death and Life of Authentic Urban Places.* New York: Oxford University Press.

Chapter 14
No-Go Areas and Chic Places: Socio-Spatial Segregation and Stigma in Guadalajara

Ulises Zarazúa Villaseñor

The Latin American metropolis is a place with a clear division among social classes being expressed in spatial terms by well-defined areas. The existence of upper- and middle-class areas, on the one hand, and of workers' districts and socially marginalized neighborhoods, on the other hand, profoundly informs the regional urban landscape. Traveling freely through a Latin American city could be equated to strolling around social differences exposed in showcases. Diverse and well-defined social milieus could be appreciated behind the display cases. Where urban segregation is highly pronounced, namely through the unequal distribution of social, economic, and cultural resources in urban space—including, among others, walls, gated communities, and no-go areas—as well as through different social sectors and ethnic groups, a radicalization of the social consequences may take place. The marked urban segregation leads to the emergence of different neighborhoods with inhabitants of very similar characteristics and thus to distinct social identities. Finally, such urban processes produce the stigmatization and social exclusion of particular social groups and the avoidance of their spaces as "no-go areas" (Guerrero 2006).

Speaking on the role of urban imageries of fear in a stratified society, Pierre Bourdieu (1993) strongly argues against the empiricist approach that tends to explain a no-go area out of itself. Contrarily to that approach, what happens in such a place is directly connected to governmental neglect and the withdrawal of the state, expressed in the absence or low quality of public services and urban infrastructure. In other words, a dangerous district should be looked at in relation to the local power. A no-go area can be explained by its role in a particular society (Bourdieu 1993). In order to break with this empiricist framing, an insightful attempt should analyze the relationships between the social and physical structures of space.

In a stratified and hierarchical society, social space is reified and expresses itself through a "naturalization effect" within the physical space in such a way that the social space and its oppositions turn out to be encrypted and engraved in the urban territory. Moreover, the social oppositions reified in the physical space tend to reproduce themselves in the local imageries and discourses through categories of identification that divide and set up an opposition: chic and "non-chic" areas, dangerous and safe districts, nice and ugly neighborhoods. In this manner, the stigma of a specific district and the urban imageries of fear attached to it can be understood as mechanisms of symbolic reproduction placed at the service of the reified social space. At the other end of the social and physical structure of space, the urban imageries constructed around upper-class residential areas and suburbs also deepen the segregation within a metropolis and naturalize social differences, as they connect upper-class neighborhoods to notions of chicness, safety, closeness to nature, and the added value of real estate. Nevertheless, beyond the urban imageries tied to upper-class districts, such areas as well as the advertising and the products related to them can be seen as part of the urban cultural economy. According to Allen Scott:

> Whatever the physico-economic constitution of such products (cultural products), the sectors that make them are all engaged in the creation of marketable outputs whose competitive qualities depend on the fact that they function at least in part as personal ornaments, modes of social display, aestheticized objects, forms of entertainment and distraction, or sources of information and self-awareness, i.e. as artifacts whose physic gratification to the consumer is high relative to utilitarian purpose. (1997):

In this chapter, I use the framing proposed by Bourdieu as well as the oppositional categories usually employed to imagine and name different parts of the city of Guadalajara, Mexico. In order to make the analysis more explicit, I chose two very different districts, located at opposite ends of the social and physical structures of space in this Mexican city. I depart from the statement that the social construction of opposing urban imageries, concerning marginalized and prestigious areas, is a symbolic mechanism of strengthening social differences and their reification within a city. Such urban imageries, as pointed out above, should be understood in relation to local society as a whole. On the other hand, although the urban imageries belong to the cultural dimension, they produce a set of effects on and empirical consequences for the realm of urban reality. The imageries, for instance, inform and even guide social practices and, above all, the use of urban space on behalf of different social actors. In this sense, some scholars argue that the urban imageries of fear can generate new paths and patterns of behavior for inhabitants of the metropolis; the fear and its imageries are structuring features of social behavior and practices in the city (Silva 1992, Guerrero 2006, Hiernaux 2007). Finally, I also approach the upper-class neighborhood of Puerta de Hierro as an excellent locus that blends style and economic power; namely a district which can be understood by the concept of urban cultural economy. Unlike the lower-class district of Santa Cecilia, Puerta de Hierro, from the very beginning of its short history, has been the subject of a deep commodification process.

The Media: Production of Urban Imageries

The urban imageries of fear and the corresponding stigmatization of the areas they are attached to are produced by the elites and echoed in the media. The mass media of Guadalajara, a highly segregated metropolis, convey a series of different urban imageries of fear concerning the lower-class neighborhoods. Such images, although symbolic constructions, are linked organically to social practices, attitudes, and even official policies, tending to stigmatize and exclude particular social groups and places. Following Pierre Bourdieu (1993) and Norbert Elias (Elias and Scotson 1965), the stigmatization of an area, a key concept in this analysis, should be understood as a two-directional process in which certain areas become nicer and chic through the demonization of others. The media gaze at no-go areas reinforces the urban segregation and urban imageries of fear, on the one hand, and the chic and safe nature of more upscale districts, on the other hand.

Some urban imageries are related to upper- and lower-class districts and displayed in the local press, a privileged locus in which the imageries, through depictions and pictures, are set in motion. Using discourse analysis (Potter 1998) and image analysis (Champagne 1993), I interpret a selection of newspaper reports, images, and advertisements related to two paradigmatic districts, located at the opposite extremes of the social and physical structures of space: Puerta de Hierro, an upper-class gated community in the western part of the city, and Santa Cecilia, a working-class neighborhood on the eastern side of Guadalajara and a stigmatized district *par excellence*. Due to the methodological approach, the characterization of both places should be understood as the two sides of the same process of differentiation that extremely stigmatizes one neighborhood and exalts the other. I used the following analytic categories:

Context   What appears to be the central information in the news items and gives the reader and idea about the district? What are the circumstances and situations that surround and call forth the appearance of the district and its people as newsworthy subjects according to the paper?

Place   How is the neighborhood depicted? Which adjectives are employed in these depictions?

Actors   Who plays the leading role? Who appears as extra? How are they portrayed? Which adjectives are related to every actor?

Photography   What and who is portrayed in the pictures? Which stories do the pictures tell us? The image, whether picture or video, exerts a powerful "evidence effect" on the readers, even more than discourse, arousing emotions and a kind of dramatization effect. When the photograph is presented as "factual evidence," it conceals the fact that it is produced by selection and construction processes (Champagne 1993).

Metanarrative   What is the larger story that can be assembled using the small stories or news items?

The sample selected for this analysis was a statistically non-representative sample of newspaper articles, encompassing 139 news items and 9 advertisements, issued on the local papers *El Informador*, *El Occidental*, *Público*, and *Mural*. The generalizability was also out of the goals of the sample design from the very beginning. However and despite its non-representativeness from the statistical point of view, the sample covers a meaningful scope of issues and topics traditionally related to the two districts.

Guadalajara: A Divided City

Guadalajara is the capital city of the state of Jalisco and, with 4.1 million inhabitants, the second largest city in Mexico. Since its foundation in 1542, the Spanish elites and the *mestizo* population settled in the western part of the town, whereas most of the indigenous people occupied the east. The San Juan de Dios River was a "natural barrier" between both sections, a barrier that even nowadays plays an important symbolic role in defining the two halves of the city. Throughout history, the local elites, the upper classes, the representatives of the local government, and the head of the Catholic Church have settled on the western side of Guadalajara. In contrast to that, the east side has been the traditional setting for workers, poor immigrants, indigenous population, and even the seat of the red-light district. The best public parks and public services are also concentrated in the west. Those differences have deepened since the 1970s and 1980s, when US-American-style shopping malls and gated communities were established on the west bank of the river.

Everyone has the right to imagine the city, but the local elites own the media and, therefore, the power to construct hegemonic narratives that displace alternative framings. In this way, old and new urban imageries have related the west side with adjectives such as "nice," "safe," "clean," and "decent," whereas East Guadalajara has been constructed as a dangerous, ugly, unsafe, dirty, immoral and indigenous area. Even though the current urban structure is more complex, and the poor districts are not exclusively concentrated on the east side, the urban imageries of the two halves of the city and the San Juan de Dios River (now Independencia Avenue) as a symbolic barrier are still powerful influences that inform identities and the everyday life of the urban population (De la Torre 2001).

The 1980s and 1990s witnessed the persistence of a strong economic crisis and the growth of the proportion of people living below the poverty line. Guadalajara, as other metropolises in the region (McLaughlin and Muncie 1999), intensified its segregation processes through a narrative of fear that highlights crime, insecurity, and immorality extended in the city. This segregation process was expressed by the quick construction of gated communities and shopping malls along the west side of town. Other practical outcomes of those imageries include the intensification of police harassment and "routine checking" of lower-class youth gangs and indigenous people, especially on the west side and downtown.

Despite the prevailing of the *mestizo* population in the city and the difficulty to find areas that directly continue the colonial ethnic segregation, where neighborhoods housed a single population group, Spanish, indigenous, or *mestizos*, the influence of the colonial times seems to still be alive today. The old "ethnic order" is still visible in Guadalajara if one travel from west to east and vice versa. The closer one gets to the east side of the city or to the poorer districts, the more frequent inhabitants are dark-skinned and indigenous. When going westwards, on the contrary, a "skin whitening" process seems to take place step by step. This ethnic order emerges unmistakably and suddenly in the cases in which maids, babysitters, construction workers, and street gardeners with an indigenous background, who work on the West side, meet on weekends for leisure in public parks or open areas in the middle of wealthy West side neighborhoods. In these instances, it has not been unusual that some of the rather light-skinned *mestizo* inhabitants have started to organize, using different resources, even the local police corps, to "drive out the invaders" and recover a bit of the dear and lost old social order (Zarazúa 2006).

The frequent police harassment against working class youth and darker-skinned people who dare to stroll in middle- and upper-class districts and wealthy malls further expresses this "out-of-order" situation. Puerta de Hierro, a western gated community, and Santa Cecilia, an eastern working-class district, are representatives of the extremes within the social and the physical structure of space of the city; a structure highly segregated and informed by issues of social class and income as well as by a more subtle ethnic order.

Media Coverage of Santa Cecilia

Despite the variety of reported subjects concerning Santa Cecilia, reports about crime and violence overwhelmingly dominate the local newspaper coverage of this neighborhood. Almost 75 per cent of the news items belong to the aforementioned topics. Besides the crime reports, Santa Cecilia also appears in the papers in combination with public investment and urban infrastructure endeavors directed at the area (urban infrastructure); the famous boxers who live there (celebrities); the plot of land endowed to the Universidad de Guadalajara by the Guadalajara County in order to build a high school, a plot that turned out to be an old landfill and was thus rejected by the university (High School Affair); the visits of Pope Jean Paul II in 1979 and of the left-wing party presidential candidate in 2006 (visitors); a series of features in which neighbors, non-governmental speakers or the paper itself denounce some failure or evaluate particular characteristics (denouncements and diagnoses); and chronicles or art-related commentaries on the district (cultural gaze).

Although a range of subjects are reported upon within the district, newspapers have clearly favored matters of crime. Such a biased angle can be observed in Table 14.1.

The news items about occurrences of crime and street violence represent an overwhelming number of the newspaper reports on Santa Cecilia, encompassing 74.1 per cent of newspaper coverage related to the district. The news reports unmistakably show a clear emphasis on seamy and violent stories about Santa Cecilia. This "favored angle" from which to observe and interpret the area, openly informs the kind of urban imageries triggered off by the very mention of the neighborhood's name. Since the foundation of Santa Cecilia and throughout its subsequent history, the local papers have favored criminal events, turning them into the most important and even "natural" framing through which to interpret the district. Santa Cecilia, as well as other lower-class neighbourhoods from eastern Guadalajara, was consistently "taking its place" inside the crime sections of the newspapers as though frequenting its *natural habitat*. This gaze has been highlighted not only by the numerical predominance of the crime-related news but also by the kind of pictures included in such publications: they are frequently eye-catching and imposing images, implying a "dramatization effect" that is clearly lacking from other kinds of newspaper photographs (Champagne 1993)[1]. In this way and insofar as the quarter was frequently featured in the crime sections of the local newspapers, Santa Cecilia gained existence and meaning as a media-constructed and symbolic place.

Table 14.1   The Newsworthiness of Santa Cecilia According to the Local Newspapers

Subject	Number of Articles	Percentage%
Crime	103	74.1
Urban Infrastructure	10	7.2
Celebrities (Boxers)	8	5.8
High School Affair	7	5.0
Visitors	5	3.6
Denouncements and Diagnoses	4	2.9
Cultural Gaze	2	1.4
Total	139	100.0

Source: 139 news items about Santa Cecilia taken from the Guadalajara newspapers *El Informador, El Occidental, Público*, and *Mural*, 1999–2007.

---

1   Due to copyright restriction the pictures can only be seen online (see References).

The Street Gangs

The gangs, especially the gangs formed by youngsters of lower-class districts, have visibly become a favored subject in the crime sections of all the papers analyzed. In the case of Santa Cecilia, the papers have symbolically constructed "the gang problem," and nowadays the gang is a favored representative of the district in the media discourse. Departing from a rather conservative "law-and-order" point of view, this social sector has been methodically related to a broad spectrum of crimes and "anti-social" practices, including street fights and shootings, robbery, burglary, drug addiction and retailing, vagrancy, graffiti spraying, homicide, manufacturing of Molotov cocktails, and even the use of female members as sexual objects. Thus, the street gangs have been targeted systematically, producing what Stanley Cohen calls "moral panic" around them (1980). A moral panic occurs when a group or social activity is perceived as a threat to the stability and well-being of society. The media attention amplifies the behavior of the groups under scrutiny. Santa Cecilia's gangs, along with youngster gangs from Guadalajara's other lower-class neighborhoods, are similarly demonized. In so doing, the local newspapers tend to echo the police's point of view and thus reproduce the perceptions of law enforcement officers and their superiors. In the case of Santa Cecilia, its gangs have been strongly related, in meaningful and vivid ways, to the urban imageries of fear associated with the district.

Depicting Through Words: Morally Loaded Terminology

I found that, depending on the newspaper section, Santa Cecilia is depicted in a more or less negative way. The usual portrayal of the neighborhood found in the crime section frequently will use negatively loaded terms, which are rather inclined to deepen the stigma of the area as a dangerous place, full of dangerous persons. The following terms are usually related to the people: "scoundrel," "rogue," "wicked man," "evil guy," "idle," "vagrant," "lay-about," "lazy," "drug addict" (crime section). Terms and phrases related to the place include "troublesome," "hot-spot," "no-go area" and "problematic area," "no man's land," "a place where the gangs rule," "at night there is a forced curfew." Therefore, the depiction in this section tends to demonize the place and its inhabitants, calling forth specific imageries of fear. Unlike the crime section, the regional section employs terms to depict the neighborhood that trigger imageries of need and lack, highlighting the "marginal nature" of the place. Especially during the 1970s and 1980s, the regional section treated the neighborhood to a certain extent in a paternalistic manner, presenting inhabitants as passive recipients of the magnanimous generosity of the governor or mayor, respectively. Ultimately, the media construction of Santa Cecilia arguably depends on the different goals and styles of the respective section.

The Metanarrative of the Wild Jungle

A metanarrative can be understood as a global narrative scheme that orders knowledge and experiences (Stephens and McCallum 1998). In other words, a metanarrative is a big story that contains and explains many little stories. The metanarrative can help us to fill in the words which can be read between lines in newspaper items, namely a sort of non-verbalized nightmare that is rarely expressed as a whole by media and outsiders' discourses. Using the metanarrative as a guide, I put together the parts, assembling the scattered single elements in order to have a more complete vision of the local urban dystopia called Santa Cecilia.

There have been other studies on the images and portraits of marginal(ized) neighborhoods conveyed by the media. Miguel Ángel Aguilar (2007) states that the metanarrative of the urban periphery as a place of chaos *par excellence* is a common image present in Latin America urban imageries. As part of such imagery, the rules and norms that control the peripheral neighborhoods seem to be different; at least from the ruling norms of downtown and affluent quarters. On the contrary, the periphery gives the impression of being structured according to an unusual order.

At the local level, Santa Cecilia has become a kind of urban symbol of "beyond the boundaries," not only geographically but also in a symbolic sense. The district has been portrayed as an "out-of-place" setting, as a place ruled by disorder: it lies in the remote far east of the city, and this remoteness somehow produces a distance in character and nature. Therefore, the geographical distance from the power centers of the elites in Guadalajara and the symbolic difference in relation to the same centers can disguise the social distance and exclusion undergone by Santa Cecilians (Bourdieu 1993). The media depictions usually highlight the distant location and the peripheral nature of the neighborhood as central features that inform it. The terms used to describe the district are not neutral words that would be generally employed, but instead look to depict a particular situation or place. "Gang member" (*pandillero*) already carries a moral disqualification and connotes fear and rejection. The media discourse does not describe the youth of poor neighborhoods who gather together to chat, play, and use drugs. Instead, they are usually portrayed as "vagrants and scoundrels" (*vagos y malvivientes*); the very usage of certain words for depicting the world implies an ideological stance and ultimately suggests the kind of attitude that the readers should adopt after reading. As stated by Jonathan Potter (1998), discourse and words create social reality; they are not simple and neutral tools for depicting the world. The common media discourse and words concerning Santa Cecilia assess and disqualify the quarter.

The Metanarrative of Lack

If Santa Cecilia is covered within a newspaper section other than the crime section, the news items usually present the neighborhood as a place in need or "lack" of something. Even though these news items do not refer exclusively to Santa Cecilia, the district, whether featured alone or as member of a "blacklist" with other neighborhoods, is systematically portrayed as a place that lacks desirable goods, services, or urban infrastructure. In this manner, it is depicted as a place constantly defined by need, deficiency, and shortage. Hence, Santa Cecilia appears as a neighborhood without sidewalks and with very neglected cobblestone streets (*El Informador*, September 9, 1972); as one of the less-wooded areas of the city (*El Informador*, June 5, 2006); as a district with no public preparatory school (*El Informador*, February 15, 2006); as one of the "most troublesome" neighborhoods in terms of the coexistence among neighbors, which the municipality cannot attend to, due to its lack of qualified personnel (*El Informador* 27.09.2006); as a quarter with no local police station (*El Informador*, February 19, 2006); as an area that borders a peripheral ring with no street lights and with many potholes (*El Informador*, January 23, 2007); as a district with no public nurseries (*El Informador*, March 15, 1999); as a neighborhood with neither a public square nor social community spaces (*El Informador*, August 18, 2000); as a district whose public playground the municipality has deeply neglected (*El Informador*, January 20, 2004); as an area with an ineffective municipal waste collection service (*El Informador*, April 21, 2005); and, finally, as a quarter whose inhabitants have a low level of education and thus face unemployment or work only in low-paid jobs (*El Informador*, March 7, 2004).

This metanarrative of absence, depicted in many news items related to the neighborhood, appears in the regional section, a newspaper section that covers a wide variety of local and regional news, ranging from economic, social, and political to cultural and artistic events. According to this discourse, the lack and absence seem to inform life in the neighborhood as a sort of inherent characteristic. The metanarrative usually depicts the lack in Santa Cecilia by contrasting the district with "somewhere else:" others places and neighborhoods of Guadalajara that are never mentioned by name. This implicit or disguised comparison, which positions the neighborhood "against" other undeclared districts, is used as a central tool in depicting and emphasizing the "lacking" nature of Santa Cecilia. Similar to the process of stigmatization, through the implicit comparison used in the metanarrative of lack, Santa Cecilia is negatively portrayed, whereas "somewhere else," which hypothetically could be localized, indeed, on the west side of the city, is implicitly presented as a better place. However, insofar as this story of lack conveys a negative image of the district, it could also be seen as another discursive element that constructs and reinforces the stigma and urban imagery of fear related to Santa Cecilia.

Media Coverage of Puerta de Hierro

Puerta de Hierro is an upper-class gated community located in Zapopan County in the Metropolitan area of Guadalajara, on the west side of the city. It is one of the most famous and representative walled residential areas of the city. Guards survey the two main entrances or gates, but the district is also subdivided into smaller gated communities, each of them with one or two more private armed guards. This pattern of gated communities inside a gated community is also found in other important local walled neighborhoods such as Las Cañadas, Valle Real, Bugambilias and Jardín Real and expresses an insatiable sense of need for greater self-protection and security.[2]

Despite its local fame, the neighborhood does not have the "media presence" that Santa Cecilia does. Most of its appearances in the local papers take the form of advertisements by Puerta de Hierro Club Residencial, a locally well-known and trademarked real estate company, as well as private classified ads in which particular luxurious mansions are put up for sale or a raffle for a "luxurious mansion" by a private university (TEC de Monterrey) is announced. Real estate advertisements use the following words and catchphrases to depict the district: "exclusive residential area," "the whole peacefulness" (quality of life); "24-hour security guards, seven days a week" (safety); "place with excellent resale value" (first-rate investment); "rich natural environment" (closeness to nature); "a new lifestyle," and "now you can be a member" (club effect).

The raffle announcements highlight the "luxurious character" of the mansions and the "exclusiveness" of the district. These paid announcements even report the blessing of the mansions by a Catholic priest. Studies on the local gated communities' advertisements point out the presence of seven main concepts: safety (Puerta de Hierro's logo is a strong and closed metal gate), exclusiveness (invoked via connotations of style and elegance as an upper-class residential area and a lifestyle affordable to only a few: exclusiveness is thus comparable to social exclusion), comfort and infrastructure (country club, tennis courts, swimming pool, golf course as symbols of status), privacy (pictures showing a single family using the public places), natural environment (however, the remoteness of the gated communities forces their inhabitants to depend on their cars, ironically reducing the alleged environmentalism), location (in a highly segregated and fragmented city it is important to be near shopping malls, private universities, main roads, and green areas), and resale value and the protection of real estate investment (Ickx 2002).

Puerta de Hierro's advertisements clearly share those concepts. Their catching phrases and images connote high status and power. Although the corresponding media coverage is less extensive than Santa Cecilia's, the announcements have frequently given the district an important place in the urban imageries of the city. Throughout the different advertisements, repeated mention is made of the "exclusive suburb" Puerta de Hierro, and readers are invited to "become a member," which exemplifies what Pierre Bourdieu (1993) calls the "club effect:" While the stigmatized district symbolically demeans its inhabitants, the chic neighborhood, established via the active exclusion of undesirable persons,

---

2   The increasing fear has even lead to people who already live in such gated labyrinths to build a *panic room*, the last fantasy of home security against burglary, kidnapping, etc.

symbolically exalts its residents, allowing them to share the capital they have collectively accumulated. The logo of Puerta de Hierro visually captures the club effect and the exclusiveness, as it features a closed iron gate (evoking hardness and impenetrability, hence security) and a fortified wall that symbolizes the separation of "inside" and "outside." Such images represent the interior space as the locus of comfort and privilege and the "outside" rest of the city, namely the space of the undesirable social strata, as an antipode to Puerta de Hierro, marked by insecurity, danger, disorder, and lack of privacy. Finally, the wall can be seen as a simultaneous symbol of exclusiveness and active social exclusion.

The inhabitants of Puerta de Hierro appear less frequently than the Santa Cecilians in the media. If they are featured, the newspapers depict them instead as victims of burglary or car theft, as an organized community rejecting the installation of a judiciary police station near the district, as happy organizers of jet set parties, and even as delinquents in the crime section.[3] Unlike the imageries triggered off by the mention of Santa Cecilia (street gangs, delinquency, drugs, and violence), the images usually related to Puerta de Hierro involve power, prestige, high status, and capital.

Becoming Chic by Close Contact

Puerta de Hierro is a common setting for high-society and local jet set events as well as a fashionable place for inaugural ceremonies of expensive upper-class commodities like cars, fine jewelry, and banking services. Luxury objects and prestigious trademark companies[4] have either offices or outlets inside the suburb or the surrounding areas. Puerta de Hierro is also a geographical—and symbolic— reference in advertisements, and regularly appears in maps of newly inaugurated businesses. The jet set parties, upper-class weddings and ceremonies, and the endless parade of First World trademarks seem to intensify the chic aura that already surrounds the district. European car dealerships, famous Spanish architectural projects, French fine jewelry exhibitions, and international golf tournaments played at the area's golf course are attracted by this aura and, with their own presence, intensify the chic nature of the district. The relationship between Puerta de Hierro and trademark businesses turned out to be mutually advantageous: The

---

3   I found three articles from the Crime Sections involving Puerta de Hierro inhabitants: one case of rich boys stealing motorbikes from another gated community and two cases of members of organized crime (auto thieves and drugs lords) living in the district. Those three criminal reports contrast sharply with the 105 criminal news items about Santa Cecilia in the same period.

4   Among the several consumer products targeted at the upper classes, I found Mercedes-Benz, Toyota, Alfa Romeo, Cartier Jewelry, Louis Vuitton, Zara and Gucci clothing and accessories, Bank of America, HSBC, International Golf Tournaments, and others.

trademarks ensure profits because of their location close to upper-class consumers, and the suburb seems to gain a kind of "First World Space" label.

The advertisement of the district and the related cultural objects show the strong relationship of Puerta de Hierro with different modes of social display and conspicuous consumption and the use of commodities as personal ornaments. The district seems to live inside the advertising; in a virtual place that endows it with glamour, sophistication, and chicness. On a very local scale, Puerta de Hierro epitomizes the place that blends notions of style and culture with the strong economic power of the native elites. Through its advertisement, Puerta de Hierro appears as a cultural product that implies the uniqueness of not being like others.

Conclusion

Santa Cecilia and Puerta de Hierro can be seen as opposite extremes of the physical and social structures of space in Guadalajara. The urban imageries of fear produced around Santa Cecilia (and the "disorder" that nowadays rules the city) are the same fears that lead the upper classes to enclose themselves in districts like Puerta de Hierro, creating a new kind of no-go area. Both spaces and their symbolic value should be understood within Guadalajara's local society. Neither Santa Cecilia nor Puerta de Hierro exist independently and cannot be explained without considering the stratified and hierarchical local society in which both districts are immersed.

While the remoteness is presented in a negative way with regard to Santa Cecilia—equating physical distance from the center with distance from law and moral order—in the case of the upper class suburb, the remoteness is presented in the advertisements as something positive and related to privacy and closeness to nature. The media approach to Santa Cecilia tends to focus on news about crime, creating metanarratives of chaos, wildness, and lack, whereas Puerta de Hierro is mostly portrayed through advertisements that emphasize values such as exclusiveness, safety, comfort, privacy, closeness to a natural environment, and an optimal location and resale value. Puerta de Hierro is also depicted in relation to countless jet set parties and upper-class trademark business ceremonies, giving the suburb a chic and attractive aura. By advertising itself as well as elegant trademarks, Puerta de Hierro, usually appearing as a setting or as geographical reference, has become an important part of the local urban cultural economies. The district appears in the newspapers as an object that can be symbolically consumed and is tied to high status, style, and economic power. Finally, the urban imageries conveyed by the local media subtly reinforce the urban segregation of the city of Guadalajara and the stigmatization of certain areas. The process of urban stigmatization, following Norbert Elias and John Scotson (1965), is one in which some districts lose and others heighten their reputation.

Works Cited

Aguilar, M. Á. 2007. *Leer e imaginar la periferia urbana: Atributos asociados con el municipio Valle de Chalco en la prensa escrita.* Paper to the Second Coloquio Internacional sobre Imaginarios Urbanos, México D.F.

Bourdieu, P. 1993. Effets de lieu, in *La misère du monde*, edited by P. Bourdieu. Paris: Seuil, 249–62.

Champagne, P. 1993. La vision médiatique, in *La misère du monde*, edited by Pierre Bourdieu. Paris: Seuil, 95–123.

Cohen, S. 1980. *Folk Devils and Moral Panics: The Creation of the Mods and the Rockers.* Oxford: Martin Robertson.

De la Torre, R. 2001. Fronteras culturales e imaginarios urbanos: la geografía moral de Guadalajara, in *El centro histórico de Guadalajara*, edited by D. Vázquez, R. De la Torre, and J. L. Cuéllar. Guadalajara: Colegio de Jalisco.

Elias, N. and Scotson, J. 1965. *The Established and the Outsiders: A Sociological Enquiry into Community Problems.* London: Cass.

Guerrero, R. M. 2006. Nosotros y los otros: segregación urbana y significado de la inseguridad en Santiago de Chile, in *Lugares e imaginarios en las metrópolis*, edited by D. Hiernaux et al. México, D.F.: UAM–Iztapalapa.

Hiernaux, D. 2007. Los imaginarios urbanos: de la teoría y los aterrizajes en los estudios urbanos. *EURE: Revista latinoamericana de estudios urbanos Regoinales*, 33(99), 17–30.

Ickx, W. 2002. Los fraccionamientos cerrados en la Zona Metropolitana de Guadalajara, in *Latinoamérica: Países abiertos, ciudades cerradas*, edited by L. F. Cabrales. Guadalajara: Universidad de Guadalajara–UNESCO. 117–41.

McLaughlin, E. and Muncie, J. 1999. Walled cities: surveillance, regulation and segregation, in *Unruly Cities? Order/Disorder*, edited by S. Pile, C. Brook, and G. Mooney. London and NewYork: Routledge, 96–136.

Potter, J. 1998. *La representación de la realidad: Discurso, retórica y construcción social.* Barcelona: Paidós.

Scott, A. J. 1997. The cultural economy of cities. *International Journal of Urban and Regional Research,* 21(2), 323–39.

Silva, A. 1992. *Imaginarios urbanos: Bogotá y São Paulo: Cultura y comunicación urbana en América Latina.* Bogotá: Tercer Mundo Editores.

Stephens, J. and McCallum, R. 1998. *Retelling Stories, Framing Culture: Traditional Story and Metanarratives in Children's Literature.* New York and London: Garland.

Zarazúa, U. 2006. *Crónicas marginales: Cinco espejos para reconstruir la ciudad.* México, D.F: UAM.

A sample of the pictures, articles and ads can be found at: http://hemeroteca. informador.com.mx/ Several dates: 19.01.72 p. 8–A, 16.10.82 p. 8–A, 11.02.94 2–A, 27.07.03 1–E, 26.07.93 5–G.

Chapter 15

Spaces of Alterity and Temporal Permanence: The Case of San Francisco's and New York's Chinatowns

Selma Siew Li Bidlingmaier

The largest Chinese community outside of Asia ... Chinatown is also the most densely populated of San Francisco's neighborhoods. Chinese laborers began coming to San Francisco in the mid-19th century as refugees from the *Opium Wars*, and were put to work constructing the railroads of the West. Racism swelled after the tracks had been laid and Gold Rush prosperity declined. In the 1880s, white Californians secured a law against further Chinese immigration to prevent the so-called "Yellow Peril". Stranded in San Francisco, Chinese-Americans *banded together* to protect themselves in this small section of downtown ... Today tourists are drawn to its alleyways and markets, which pulse with the *sights and smells of Chinese-American culture* ... today's Chinatown is one that *celebrates Chinese-American culture. Chinatown residents caution against mistaking Chinatown for a re-creation of China (if anything, Chinatown is more of a snapshot of Chinese life forty years ago, preserved by immigrants who remember that China).*

(*Let's Go San Francisco* 2003: 9, emphasis mine)

This description of San Francisco's Chinatown from the 2003 edition of the popular travel guide *Let's Go* succinctly captures two popular, interrelated conceptions of Chinatown. First, it is a space of absolute otherness. The word "China-Town" itself distinctly delineates the space as geographically, nationally, culturally, and ethnically dissimilar in comparison to the wider, normative spaces of unmarked "American-Town." The enclave is placed in an ambiguous position of being within the national boundaries of the United States, yet at the same time a figment of imagination and "if anything" a representation of China "forty years ago." The abstractness of this spatial configuration becomes a tangible reality as tourists venturing past the infamous arches and gates of Chinatowns across the country, armed with their cameras, expectantly believe that they are entering another world marked by the "sights and smells of Chinese-American culture." Besides Chinatown's supposed alterity, the excerpt above also articulates another concept of its almost-magical quality—temporal permanence. Chinatowns are often imagined to be spaces where time is suspended, conserving cultural authenticity, traditions, and all cultural artifacts. The space is made to function as an ethnographic museum which systematically and coherently codes, classifies, and constitutes

Figure 15.1   Dragon's Arch, Grant Street, San Francisco. Photograph by the author

cultural difference. These two attributes of Chinatown (its prescribed alterity and timelessness) render the space a permanent fixture of otherness within the cityscapes of the United States. Moreover, these attributes stabilize, purify, and de-politicize the myths, stereotypes, and popular imagination of both Chinese Americans and the spaces they inhabit.

Contrary to the way Chinatowns have been represented and perceived as marginalized spaces at the fringes of society, the enclaves have never been detached from or unaffected by the wider society. Rather, Chinatowns have been and still are intricately connected to various sites beyond both imaginative and institutionalized borders. With reference to Michel Foucault's concept of heterotopia, Chinatowns are related to "all the other sites but in such a way as to suspect, neutralize, or invert the sets of relations that they happen to designate, mirror, or reflect" (Foucault 1986: 24). For the purpose of this paper, I will only concentrate on its relation to "American" sites. I intend to locate these "othered" spaces within the dialectics of Foucault's philosophical musings on heterotopia. I will propose a means to answer two questions: 1) Why is Chinatown conceived as an "othered" space and what are its myths? And, 2) how and why have the technologies of myth-making sustained and maintained the conception of this space so successfully that these myths continue to proliferate our media and imaginations in the twenty-first century?

In the nineteenth century, rhetoric such as "densely populated neighborhoods," "opium," and "banded together" were intertwined with the discourses of social reform, health reform, and political and immigration policies. At a time of mass migration, urbanization, industrial expansion, and the increase of poverty-stricken slums, the dominant class saw the vast numbers of Chinese immigrants entering the young nation as a threat to their social order and values, as well as a competition in the labor markets. In San Francisco, Chinatown swelled as the population of Chinese immigrants quadrupled.[1] The 1880 federal census shows that Chinese immigrants in San Francisco stood at nine percent of the entire population of San Francisco. According to historian Nayan Shah, Chinatown's borders "abutted four defined zones of the city: the elite residential district of Nob Hill ... the main commercial and business districts on the south and east, and the Latin Quarter" (2001: 25). Chinatown's increasing proximity to the upper- and middle-class spaces was deemed to be a great threat. Furthermore, on the East Coast , four out of five of Chinese immigrants in New York were residing outside the boundaries of Chinatown according to the 1880 census (Tchen 1999: 248). Working as house servants and laundrymen, their presence in middle- and upper-class residential areas was greatly felt. The growth of the Chinese population in metropolises such as New York and San Francisco coincided with the emergence of a new middle class that perceived the Chinese immigrants, sorely visible amongst the

---

1   The population of San Francisco's Chinatown "climbed rapidly in the 1860s. The total population of San Francisco nearly tripled in size to 149,473, and the Chinese population quadrupled to over 12,000 in 1870. By the 1880 census, the Chinese population stood at 21,745 out of a total of 233,979" (Shah 2001: 25).

crowd of new immigrants, as vastly different and ultimately inassimilable. Their customs and traditional practices conjured up an opposing vision to the clean, healthy, moral, industrious, godly, middle-class US-American society. The fear of the "Oriental" mongrelization of US-American values, traditions, and culture was manifested in various discourses ranging from politics to medicine. The imminent but yet intangible "yellow peril" was projected onto the material spaces of Chinatown, thereby making the threat a physical, perceptible, and containable reality. By setting apart these racialized spaces, Chinatown became a site that was an inversion of the normative spaces of nineteenth-century America, understood to be detrimental to the self-perception of the young United States in its course of nation-building.

The "excess"[2] of Chinese immigrants in the mid-nineteenth century spurred ferocious discussions and investigation of the "Chinese Question" and the spaces which they inhabited. Travel writers, health officers, physicians and journalists perpetuated racialized "medical knowledge" that pronounced the Chinese a race of rat-eaters, unhygienic, and inevitably carriers of diseases such as cholera and smallpox. In a report by the *Daily Alta California* dating from 1850, editors warned their readers in San Francisco of the potential outbreak of cholera in the "filthy localities like the Chinese quarters" because "cholera delights in filth, in decaying garbage, stagnant water, as well as dirty clothing and filthy bodies: particularly when all of these are united in crowded localities" (qtd. in Shah 2001: 21). Moreover, zealous city health officials such as Dr. Thomas Logan claimed that "heredity vices" or "engrafted peculiarities" preordained the Chinese to chronic and unusual illness (qtd. in Shah 2001: 29). The popularized medical knowledge of Chinatown as a space of contamination and danger justified racist policies such as the implementation of medical taxes only applicable to the Chinese.[3] Even more significantly, it resulted in the segregation of healthcare—Chinese were denied medical care and treatment at public facilities. The threat of the diseased Chinese was perceived to be so imminent that in 1885, the mayor commissioned a cartographical documentation of the Chinese enclave that gave a detailed location of buildings designated to be Chinese restaurants, gambling houses, laundries, brothels, etc. This map, completed in 1885 and titled *Official Map of "Chinatown" in San Francisco*, became the authoritative reference for social and political agencies. This systematic mapping of Chinatown functioned not merely to identify spaces occupied by the Chinese; it demarcated the areas that were perceived to be dangerous to public health, a place outside and "other" from San Francisco.

All this coincided with the rapid modernization of medical science and the heightened obsession of sanitation and hygienic practices in the mid- to late-

---

2   The word "excess" was commonly used in socio-political and medical discourses of the day to describe the Chinese (Shah 2001, 22).

3   "In 1852, the state legislature passed a tax on passengers arriving at the port of San Francisco in order to pay the cost of the State Marine Hospital of San Francisco and public hospitals in Sacramento and Stockton" (Shah 2001, 24).

nineteenth century. From Louis Pasteur's discovery of microbes that were responsible for silkworm disease in 1864 to Robert Koch's demonstration that diseases could be identified by their specific causative bacillus in 1876 (Sivulka 2001: 60), scientific advancement ushered in a new paradigm of medical knowledge that had tremendous influence on the institutionalization of sociopolitical policies. The awareness of contamination and disease was propagated and intensified with the rise of the mass market's promotion of industrial commodities such as soap, detergent, and home sanitary utilities like bathtubs and showers.[4] Women's home journals proliferated ideals of cleanliness and hygiene to be practiced in upper- and middle-class homes, and these power/knowledges produced reverberated within discourses calling for a cleaner and healthier US-American society. Chinatown, with its dense and filthy population, was the undomesticated antithesis of the new and improved sanitized United States.

Although burgeoning metropolises such as New York were no stranger to the mounting problems of slums resulting from mass immigration, Chinatowns, unlike other ethnic enclaves, were deemed unreformable. In 1890 Jacob Riis, an immigrant himself, published his influential sociological study *How the Other Half Lives*, which documented New York's tenements at the turn of the twentieth century. He hoped that his work would make the US-American public aware of the struggles and woes of the new immigrants stricken by poverty. However, influenced by the overarching discourse of Social Darwinism, Riis declared Chinatown unreformable. Unlike the Italian Quarter, Jewtown, and the other racialized residential spaces in New York whose tragic fate was attributed to extreme poverty, Chinatown could not be reformed as the "Chinaman" was predisposed to certain traits that rendered him an inadvertent victim to barbaric behavior, unhygienic practices, as well as damaging habits like gambling and smoking opium. Riis concluded his study of Chinatown with these haunting words: "[T]here is neither hope nor recovery; nothing but death—moral, mental, and physical death" (1996: 128). So severe was the perceived threat of the Chinese to the normative order of Victorian spaces that the 1882 Chinese Exclusion Act, which denied entry to Chinese laborers, claimed that their entry would "endangered the good order of certain localities within the territory."

Despite warnings from health officials and social reformers, tourists and bohemian artists frequently transgressed the borders of New York's and San Francisco's Chinatowns. Lured by its outlandish oriental curiosities, the thrill of estrangement, the excitement of lurking danger, and exotic romance, cultural historian Emma J. Teng astutely observes that San Francisco's Chinatown "was a type of amusement park for the bohemian *flâneur*, that particular breed of walker/ voyeur" (2002: 59). Known for its prostitution houses, opium dens, gambling parlors, and mysterious legendary labyrinths, Chinatown's "spaces of unreason"

---

4   In her book *Stronger Than Dirt* (2001), Juliann Sivulka reveals a very interesting cultural history of the advertising of personal hygiene, which traces how cleanliness became synonymous with middle-class morality and civility in the late nineteenth century.

provided an escape from the rational, orderly, and oftentimes repressive "spaces of reason" of Victorian, bourgeois US-America. The appeal to enter this "space of unreason" evokes a sense of estrangement from the familiar and *heimliche* or "homey" spaces of normative middle-class society. With its narrow and dark alleys, unintelligible signs, foreign-looking inhabitants, Chinatown was *unheimlich*, "un-homey" to white Americans, both culturally as well as spatially (Teng 2002: 56). Teng has argues that Chinatown was an uncanny space as it represented Freud's definition of the uncanny as marking the "return of the repressed"—as "that which ought to have remained hidden but has come to light" (cf. Freud 1955: 225). It was a place associated with activities white bourgeois sought to keep hidden. Spatially, it was the part of the city that became a symbol for the "darkest aspect of human nature that lies submerged beneath the surface of civilized society" (Teng 2002: 56). Developing this idea, I argue that the sensation of *das Unheimliche*, or the uncanny, stems from a more profound perception of spatial alterity. Foucault describes heterotopias as mirrors which "exert a sort of counteraction on the position that [one] occupies" (1986: 24). He goes on to explain the function of a heterotopia as such:

> [I]t makes this place that I occupy at the moment when I look at myself in the glass at once absolutely real, connected with all the space that surrounds it, and absolutely unreal, since in order to be perceived it has to pass through this virtual point which is over there [in the mirror]. (1986: 24)

Chinatowns are heterotopias in that they simultaneously reflect and affirm the "realities" of the normative spaces that one inhabits, on the one hand, and at the same time force the person who encounters this othered space to a disconcerting suspicion that the spaces which he or she inhabits are but an illusion, dependent upon their definition of the other.

Although the socio-political and, to some extent, the cultural rhetoric of Chinatown has changed, it is intriguing that the imagination of Chinatowns in the twenty-first century as isolated, separate spaces within city centers such as New York and San Francisco diverges little from that of the nineteenth century. The myths associated with these spaces continue to render Chinatown exotic, dangerous, uncanny, and perpetually other. These myths emerge within various discourses today ranging from tourism to economics. What is most intriguing is why and how these myths have been sustained over more than a century. In *Of Other Spaces* Foucault explains that one principle aspect of heterotopias is that they are "often linked to slices of time" (1986: 26). He goes on to name two main types of heterotopias: heterotopias of indefinitely accumulating time and transitory heterotopias. According to Foucault, the museum and the library best exemplify heterotopias of indefinitely accumulating time. These spaces are a type of general archive, "enclosed in one place all times, all epochs, all forms, all tastes, the idea of constituting a place of all times that is itself outside of time and inaccessible to its ravages" (1986: 26). Chinatowns are believed to be archives of authentic

Chinese culture, a place to visit, to experience and learn about the Orient within the comfort zones of the Occident. It is a space that authenticates and reinforces the idea of the cultural and social normality of "American" sites and identity, while at the same time fixing, stabilizing, and attaching these myths to spaces represented as unchanging and ageless. There is a wonderful series of postcards sold in the tourist stores of San Francisco's Chinatown entitled *Chinatown Then and Now*. One of them juxtaposes two pictures taken at the intersection of Grant Avenue and California Street—one black and white picture taken in the 1960s, and the other taken in recent times. The caption "Then and Now" calls to attention the stunning similarities of the landscape and the buildings, and captures an uncanny sense of "timelessness." From Edward Said's perspective, museums function as hegemonic technologies that have maintained the idea of the Orient. They function to establish socio-political authority and superiority by establishing power/ knowledge paradigms to speak of and display the other (1994: 1–27). Imagined and represented as Oriental museums of the West which have meticulously preserved Chinese culture, Chinatowns become a consistent referential space that maintains a "system of knowledge about the Orient" (1994: 6).

The durable and relatively consistent stereotypes, perceptions, and imagination of Chinatown are maintained through the hegemonic discourses inscribed onto the space as well as through the *performance* of otherness. Ruminating on how stable core gender identities are formed, Judith Butler asserts that gender has to be repeatedly performed. Performativity according to her is "the reiterative and citational practice by which discourse produces the effects that it names" (Butler 1993: 2). Chinatowns are performative spaces. Tourists entering Chinatown often come with preconceived ideas of what to expect, what to eat, and what to take home as souvenirs. They are actors who, by performing the script regulated by the various discourses, engage in affirming and perpetuating spatial otherness. From the moment the cultural tourists traverse the flamboyant arches of Chinatown to the moment they leave, Chinatown becomes a symbolic space in which cultural and spatial differences are practiced, staged, and reinforced.

On a more subversive level, Chinatowns are very much commodified ethnic spaces and have been since the nineteenth century. The seeming "timelessness" of this enclave helps to sell Chinatown by creating a consistent product for the consumption of cultural tourists. Perpetuating this cycle, these expectations of difference staged and performed by the various actors working as tour guides (there is a tour company based in San Francisco's Chinatown that offers ghost tours), museum curators, street vendors, waiters and waitresses, etc., reinforce the seemingly immutable nature of this othered space through the reiteration and referencing of the discourses of Chinatown. However, this must not be mistaken for simple deception, but rather an act of mimicry that "emerges as the representation of a difference that is itself a process of disavowal" (Bhabha 1994: 122). A wonderful example of this can be found in the act of naming a bubble-tea café in New York's Chinatown *Big Bubble in Little China*. At the onset, the name seems almost to pay tribute to John Carpenter's movie *Big*

Figure 15.2   A bubble-tea café in the heart of New York's Chinatown that satirizes John Carpenter's *Big Trouble in Little China* (1986), shot in San Francisco's Chinatown. Photograph by the author

Figure 15.3 Souvenir stores that "dot" Chinatown play on the expectations of cultural voyeurs. Photograph by the author

Trouble in Little China (1984), which perpetuates discourses of difference and otherness conceding to the hegemony of Western popular culture. In this case, the name of this café sells, as it is a normalized cultural signifier that almost all tourists would be able to read. At the same time, the name can also be read as a performance of mimicry and an overt mockery that frustrates any attempt to misread it as an allusion to difference.

The word "Chinatown" today is often used as a metaphor for a world of inscrutable otherness that is closed, exotic, dangerous, and mystical. Conversely, Chinatowns since the nineteenth century have functioned as crucial sites in the building of a middle-class American spatial identity, and at the same time continuously disrupt the binaries of East and West, Orient and Occident, by mirroring, inverting, and contesting what is believed to be the natural order of spaces. Chinatowns function as heterotopias that destabilize and un-ground the belief that spaces are self-contained and distinct from one another while demonstrating how spaces are intricately connected and related to each other. The discursive processes that construct US-American Chinatowns as spaces of otherness necessitates an illusionary temporal permanence creating a space that is seemingly "timeless"—a museum of otherness.

Works Cited

Bhabha, H. 1994. *The Location of Culture*. New York: Routledge.

Butler, J. 1993. *Bodies That Matter: On the Discursive Limits of "Sex."* London: Routledge.

Foucault, M. 1986. Of other spaces, translated by Jay Miskowiec. *Diacritics* 16(1), 22–7.

Freud, S. 1955 [1919]. The "uncanny," in *The Standard Edition of the Complete Psychological Works of Sigmund Freud*, vol. 17, edited by J. Strachey. London: Hogarth Press, 217–256.

Lui, Mary Ting Yi. 2005. *The Chinatown Trunk Mystery. Murder, Miscegenation, and Other Dangerous Encounters in Turn-of-the-Century New York City*. New Jersey: Princeton University Press.

Riis, J. (1890) 1996. *How the Other Half Lives*, edited by D. Leviatin. New York: Bedford Books.

Nwandu, A. (ed.) 2003. *Let's Go City Guide: San Francisco*. New York: St. Martin's Press.

Said, E. (1978) 1994. *Orientalism*. New York: Vintage.

Shah, N. 2001. *Contagious Divides: Epidemics and Race in San Francisco's Chinatown*. Berkeley: University of California Press.

Sivulka, J. 2001. *Stronger Than Dirt. A Cultural History of Advertising Personal Hygiene in America, 1875 to 1940*. New York: Humanity Books.

Tchen, J. 1999. *New York before Chinatown: Orientalism and the Shaping of American Culture 1776–1882*. Baltimore: John Hopkins University Press.

Teng, E. 2002. Artifacts of a lost city, in *Re-Collecting Early Asian America. Essays in Cultural History*, edited by J. Lee, I. Lim, and Y. Matsukawa. Philadelphia: Temple University Press, 54–77.

Index